Theoretical Biology

Series Editor

Yoh Iwasa, Kyushu University, Fukuoka, Japan

The "Theoretical Biology" series publishes volumes on all aspects of life sciences research for which a mathematical or computational approach can offer the appropriate methods to deepen our knowledge and insight.

Topics covered include: cell and molecular biology, genetics, developmental biology, evolutionary biology, behavior sciences, species diversity, population ecology, chronobiology, bioinformatics, immunology, neuroscience, agricultural science, and medicine.

The main focus of the series is on the biological phenomena whereas mathematics or informatics contribute the adequate tools. Target audience is researchers and graduate students in biology and other related areas who are interested in using mathematical techniques or computer simulations to understand biological processes and mathematicians who want to learn what are the questions biologists like to know using diverse mathematical tools.

Makoto Sato

Getting Started in Mathematical Life Sciences

From MATLAB Programming to Computer Simulations

 Springer

Makoto Sato
Institute for Frontier Science Initiative
Kanazawa University
Ishikawa, Kanazawa, Japan

ISSN 2522-0438 ISSN 2522-0446 (electronic)
Theoretical Biology
ISBN 978-981-19-8259-0 ISBN 978-981-19-8257-6 (eBook)
https://doi.org/10.1007/978-981-19-8257-6

This Springer imprint is published by the registered company Springer Nature Singapore Pte Ltd.
The registered company address is: 152 Beach Road, #21-01/04 Gateway East, Singapore 189721,
Singapore

Preface

In the fields of life sciences such as molecular biology, developmental biology, cell biology, and neuroscience, there is an increasing need for knowledge and skills in mathematical and computer sciences, including not only statistical analysis of data but also mathematical modeling, numerical simulation, and image analysis. In the fields of physics and engineering, it is common to use mathematical models that replace the target phenomena with differential equations in order to predict new phenomena and clarify the mechanisms behind them.

In order to understand and promote this kind of research, it is necessary to learn computer programming, but many life science researchers and students have not had the opportunity to receive such education, or have difficulty getting started due to a sense of weakness. I am a life scientist myself, and I have only done programming as a hobby, but recent programming languages are very easy to use, and computers are becoming more sophisticated, so even laymen can create practical programs. Performing simulations is almost the same as creating a program to solve differential equations, but to do so, it is necessary to learn mathematics properly. Since it is very difficult to learn both programming and mathematics, the primary goal of this book is to enable you to program without going into the details of mathematics as much as possible. It is recommended to have a good knowledge of high-school level mathematics and physics, but matrices and differential equations, which may not be taught in high school, are explained as necessary.

There are many programming languages in the world, and this book deals with MATLAB, which is sold by MathWorks. Of course, each language has its own characteristics, and in the case of MATLAB, one of its major advantages is that it is good at handling matrices. Graphing data is easy, so you can create nice looking graphs with just one line of instructions. The problem is that MATLAB requires a fee, and even if you have created a program, you cannot use it unless you have a computer with MATLAB installed. MATLAB is available at academic and student prices, and there is also an inexpensive home license called MATLAB Home. You can also use a free trial version for one month or use GNU Octave (called Octave from now on), a free programming language compatible with MATLAB.

If you are interested, why don't you install MATLAB or Octave on your PC and start right away? It may seem a little difficult, but if you find it surprisingly simple, that alone is a big step forward. You may be able to understand the contents of mathematical presentations without being afraid of them. Although the scope of this book is limited, we hope that this book will help you acquire more specialized skills, which will lead to the advancement of life science in the near future. We hope that you will take up the challenge.

Kanazawa, Japan   Makoto Sato

Contents

Chapter 1
Preparation

1.1 What Is Programming?

You might think that the CPU (short for central processing unit), which is the brain of modern computers and smartphones, would be very smart, but actually it is not. CPUs can only perform very simple calculations at the level of addition and subtraction, but by combining simple calculations and executing them at very high speeds, they appear to be smarter on the surface. In order to do this, it is necessary for humans to create a program for the computer to execute using a programming language. There are a variety of programming languages, but the language that the CPU can understand directly is the most basic one, called machine language. However, since the CPU can only understand a very limited number of things, it is very difficult for humans to program in machine language. Therefore, we generally program in high-level languages, which are easier for humans to understand. There are various types of high-level languages, but the most commonly used languages can be roughly divided into two types: compilers and interpreters. Programs written for compilers need to be translated (compiled) into machine language once and then executed, which takes a little more time, but the execution speed of the program is faster. On the other hand, programs written for the interpreter are executed by translating each instruction one by one. **MATLAB** is an interpreted language, so it is easy to use even for beginners.

1.2 Installation

First of all, let's install MATLAB; the free evaluation version of MATLAB is available for free for 30 days, so it is a good idea to install it from the MathWorks website. If you are a student, you can purchase it at a student price if you follow the prescribed procedure at a university coop, etc. The price of MATLAB alone is about

50 USD (in Japan), which is enough for a student to purchase. Why don't you take this opportunity to buy one?

Alternatively, you can install and use **Octave**, a free language compatible with MATLAB. You can download it from the GNU Octave home page (https://www.gnu.org/software/octave/) or (https://wiki.octave.org/Octave_for_macOS) if you are on macOS. After installation, you will be able to use either the command line version (CLI) or the graphic user interface version (GUI), but I think the GUI version is easier to understand because it is more similar to MATLAB. The layout is a little different from MATLAB, but there are no major differences.

1.3 Trying to Use MATLAB as a Functional Calculator

1.3.1 Calculating in the Command Window

When MATLAB is successfully installed and started, you will see the screen as shown in Fig. 1.1. When you install MATLAB, a folder named "MATLAB" will be created, and this will be the current folder. First, click on this "Command Window." You will see a blinking vertical bar to the right of ">>" (Fig. 1.1). This is called the cursor, and it means that you are ready to enter text from the computer keyboard.

In the upper right corner of the screen, there is a box with a magnifying glass. If you enter a command name or a word here that you do not understand, you can search for the related document in the official documentation called MATLAB documentation. Use this feature when you need detailed explanations. Octave has a slightly different layout but is basically the same (Fig. 1.2).

Since MATLAB and Octave are interpreted languages, the user does not have to consciously translate the instructions. Therefore, the user can execute the instructions without writing a program. For example, you can make it calculate or draw a graph as if it were a functional calculator. First, let's try using it like a functional calculator.

In the command window, let's type

1+1

and press the Enter key. Then

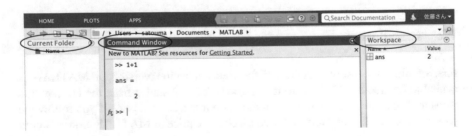

Fig. 1.1 Screenshot of MATLAB

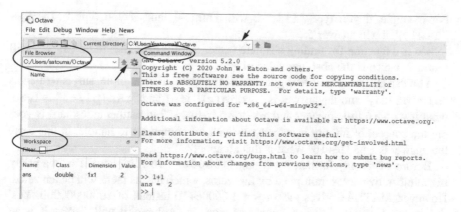

Fig. 1.2 Screenshot of Octave

```
ans =
2
```

This means that the program calculated $1 + 1 = 2$. Of course, you can also subtract and handle not only integers but also decimal points. Let's type

```
4.1-5.2
```

and press Enter:

```
Ans =
-1.1000
```

will be displayed.

Multiplication and division are different from common mathematical symbols, using "*" for multiplication and "/" for division. `3.2*1.5` will give you the answer 4.8000, and `3.14/2` will give you 1.5700.

What happens if you enter "`2^3`"? In this case, the answer is 8. Yes, "^" means to the power of. In mathematics, pi is written as π, but since there is no π on the computer keyboard, MATLAB uses the word "pi." If you type `pi` and press Enter, you will see 3.1416. In this chapter, please always type in the command window and press Enter.

1.3.2 Calculating Functions

You can also calculate square roots. The root symbol is also not on the keyboard, but you can use a function called *sqrt*. The function here is similar to a function in mathematics, which uses the number in parentheses to calculate something. In other words, if you type "`sqrt(5)`" and press the Enter key, you will get the answer 2.2361. Of course, you can put any number you want in the parentheses.

What happens if you try "`sqrt(-2)`"? Surprisingly, it returns the answer as "0.0000+1.4142i." Of course, this "i" is an imaginary number, and the square of i is −1, so MATLAB can handle imaginary numbers without any special instruction.

For example, if you type "i*i," it will return –1, and if you type "(1+2i)* (1-2i)," it will answer 5.

abs is a function for finding absolute values, so, for example, "abs(3-5)" will return 2, which is the absolute value of –2.

You can also use all other trigonometric functions. For example, $\sin\frac{\pi}{2}$ is written as "sin(pi/2)," and the answer is 1. "cos(π)" is –1 and "tan(pi/4)" is 1. So, what if we are a little meaner and ask "tan(pi/2)"? In mathematics, $\tan\frac{\pi}{2}$ is +∞, but the concept of infinity is actually difficult for computers to handle. In this case, the answer is 1.6331e+16. What is this all about?

Before we go any further, try typing "100000*100000" here. That's a pretty big number, five zeros multiplied by five zeros, which means there will be ten zeros. However, MATLAB gives the answer 1.0000e+10 instead of 10000000000. Yes, this means 1.0×10^{10}. If the number of digits is too large, it will display it as an exponent like this. Although tan(π/2) is supposed to be +∞, the computer cannot handle the concept of ∞, so it returns a very large value of 1.6331×10^{16} as the answer. Note that pi is essentially an irrational number, and there is an infinite sequence of numbers after 3.14, but the computer cannot handle such an infinite sequence of numbers and will only calculate up to a finite number of digits. Therefore, pi/2 itself in the parentheses of tan(pi/2) is not exactly pi/2.

In Chap. 3 and later, we will simulate differential equations by solving them, and in order to calculate the derivative, we need to deal with the concept of limit. But computers cannot handle infinity or infinite sequences of numbers, so we need to devise various methods.

1.3.3 Calculating Inverse Functions

Returning to functions, we can also use the inverse functions of the trigonometric functions mentioned earlier. For example, the inverse function of *sin* is called *asin* (arcsine). In other words, if "x=sin(t)," you can replace x and t to get "t=asin(x)." For example, if you enter "asin(1)," you will get the result 1.5708. If we divide this by pi and set "asin(1)/pi," we get an answer of 0.5. This means that asin(1) =π/2, and since $\sin\frac{\pi}{2}=1$, we can confirm that this is the correct answer. Similarly, the inverse functions of *cos* and *tan* are *acos* and *atan*, respectively, so type "acos (1/2)" or "atan(1)" and see what you get.

It may not be clear at this point what the inverse of the trigonometric function is useful for in the life sciences, but, for example, if you know the *x–y* coordinates of the position of a cell, and you want to express the position of that cell in polar coordinates, you can use *atan* to find the angle. A concrete example of its use is given in Sect. 2.12.

There are many other useful functions that can be used, but it is beyond the scope of this book to explain all the functions that can be used in MATLAB, so it would be better to search the MATLAB documentation (Fig. 1.1), other books, or the Internet.

You can find detailed explanations of the various functions on the MathWorks website (https://jp.mathworks.com/help/matlab/referencelist.html).

1.4 Variables

1.4.1 What Are Variables?

Next, let's talk about variables, a concept that is essential for learning programming. Basically, they are the same as variables in mathematics. For example, when you write "X=2," it means that X is equal to 2 in mathematics. In programming, however, the meaning of "=" is slightly different. Here, it does not mean equality but rather **the assignment of the right side to the left side**. For example, put the number "1" into the box labeled "X" (Fig. 1.3a). Then, unless you do something to the box "X," the number "1" will be in the box.

In fact, if you type "X=1" and press the Enter key, you will see in the command window

X =

1

In addition, if you look at the workspace, you will see that there is a variable named X in the name field, and its value is displayed as 1.

If you type only "X" again and press Enter, you will see

X =

1

MATLAB is designed to display the answer to the input command, so whether you type "X=1" or just "X", it will display the answer. But now, type

X=3;

and press Enter. This semicolon (;) is important. In this case, nothing is displayed in the command window after the line break, but you can see that the value of X has changed to 3 in the workspace. In other words, the semicolon means **"do not display the result in the command window."** Please keep this in mind as it will be important later.

Now, X has a value of 3. Type

X=X+1;

Fig. 1.3 Intuitive understanding of scalar variables. (**a**) "1" is assigned to "X". (**b**) When "X=3", "X+1" is assigned to "X". (**c**) When "X=4", "πX" is assigned to "Y". Then, "cosY" is assigned to "Z"

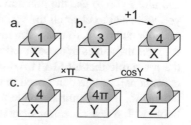

and press Enter. This is a mathematically unintelligible command, but in this case, it means "**assign the value of X+1 to X**" that is, "**put the value that was originally in the box X plus 1 into the same box X**" (Fig. 1.3b). The operation of adding a certain value to a variable and assigning it to the original variable, such as "X=X+1" or "I=I+3," is very common in programming.

Now, another common operation is to use the value of a variable to perform a calculation and then assign the result to another variable. For example, here we have

Y=X*pi;

you will see the value of Y displayed as 12.5664 in the workspace. This is the value of 4π (Fig. 1.3c). If you simply type "Y" and press Enter, you will see Y=12.5664 in the command window. Also, if you type

Z=cos(Y);

the value of Z will be cos4π, that is, 1, and typing "Z" and pressing Enter will display "Z=1" (Fig. 1.3c).

Finally, by adding a semicolon (;), it is possible to execute multiple instructions in one line. For example, "X=3; Y=X*pi; Z=cos(Y)" you can execute three instructions at once in one line.

1.4.2   Names of Variables

The name of the variable can be basically whatever you want it to be, but it cannot contain special characters or symbols. You can include numbers, but the first character must be a letter of the alphabet. The alphabet can be upper- or lowercase as follows:

xyz

a100

abcXYZ

ABCxyz

You can type "abcXYZ=1; ABCxyz=2;" and then "abcXYZ" and "ABCxyz" to see the contents of these variables (don't forget to press Enter).

However, it is not possible to use variables with the same name as functions already built into MATLAB. For example, you can't use *sqrt*, *abs*, *sin*, etc., which have been mentioned in the past. If you make a mistake and type "sin=1;," you will not be able to use the *sin* function. In such a case, type "clear" in the command window. This will reset the variable information stored in the workspace and allow you to use the function *sin* again. There will be times in the future when you will need to clear all the variables, but in that case, use this *clear* command.

Now, there are so many functions in MATLAB that it is troublesome to make sure that the names of these functions and variables do not overlap, isn't it? Actually, there is a rule that all MATLAB functions have lowercase letters (though there are exceptions), so you don't need to worry about it if your variable contains uppercase letters. So, it is recommended to use variables that contain uppercase letters. In this book, with a few exceptions, variable names begin with a capital letter.

1.5   Vectors

1.5.1   What Are Vectors?

So far, we have only assigned one number (scalar) to one variable, but just like in mathematics, we can assign multiple numbers to one variable. In other words, variables can be treated as vectors or matrices. This is actually one of the strengths of MATLAB, but let's start with vectors. Type

```
A=1:10
```

in the command window, and press Enter. Omit the semicolon (;) here. Then, we get

```
A =
1 2 3 4 5 6 7 8 9 10
```

In the workspace, the value of A is [1,2,3,4,5,6,7,8,9,10]. You can think of A as a box divided into 10 parts (Fig. 1.4). The nice thing about MATLAB is that you can treat this vector as if it were a scalar variable with only one number. For example, type

```
A*2
```

and press Enter. Again, leave out the semicolon (;) to see the result of the calculation. The result will be displayed as

```
ans =
2 4 6 8 10 12 14 16 18 20
```

The result of multiplying each of the 10 elements of vector A by 2 is now displayed. The same applies to other arithmetic operations, and you can also assign the result of the operation to other variables.

```
B=A/3;
```

The result of dividing the elements of A by 3 will be assigned to the new vector B (Fig. 1.4). In this case, there is a semicolon and the result of the calculation of B is not displayed, so we can use

```
B
```

and then press Enter without the semicolon. You will get

```
B =
Columns 1 to 9
```

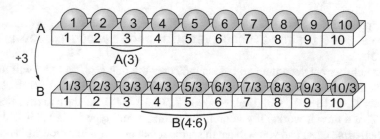

Fig. 1.4 Intuitive understanding of vectors

```
   0.3333 0.6667 1.0000 1.3333 1.6667 2.0000 2.3333 2.6667
3.0000
   10 columns
   3.3333
```

In this way, when the number of elements is large, a new line will be displayed every 10 columns. Note that we call the number of elements columns. A vector is a kind of matrix that has only one row (or one column). This will be easier to understand when we explain matrices in Sect. 1.7.

As you can see, it is very convenient to perform operations on many elements at once. For example, if you type "sqrt (A)," the square root of each element of A will be displayed.

To find the sum of the elements of a vector, use *sum.* "sum (A)" or "sum (B)" will display the sum of all the elements of vector A or B.

1.5.2   Manipulating Elements of Vectors

While it is good to be able to operate on many elements at once, how can we extract only the necessary elements from them? Here, we have

   A(3)

and press Enter. We get

   ans =

   3

In other words, the third element (third column) of vector A, 3, is now displayed (Fig. 1.4). Similarly, typing "B (5)" will display 1.6667, the fifth element of B. It is also possible to retrieve multiple elements simultaneously.

If we type "B (4:6)" and press Enter, we get

   ans =

   1.3333 1.6667 2.0000

In other words, we have extracted the fourth through sixth elements of B (Fig. 1.4).

This notation using a colon (:) is very common in MATLAB. Similar to the above, "1:5" generates a vector of integers from 1 to 5, that is, 1 2 3 4 5. Type

   A=1:0.5:5;

   A

and press Enter. Then

   A = 1.0000 1.5000 2.0000 2.5000 3.0000 3.5000 4.0000 4.5000
5.0000

The number between the two colons indicates the tick size. "A (1)" will display 1.0000, the first element of vector A. "A (5)" will display 3.0000, the fifth element of vector A. "A (3:5)" will display 2.0000 2.5000 3.0000, the third to fifth elements of A. That's how it works. If you omit the numbers and type "A (:)," it will mean **"all elements"** of A, and you will get the same result as if you just type "A" and press Enter.

Of course, you can also perform operations on this vector A, such as "B=A*pi;" and "C=sin(B);" to check the contents of B and C.

1.6 Drawing a Graph Using a Vector

1.6.1 Using plot Function

In MATLAB, you cannot only handle vectors easily but also display the contents of vectors as graphs along the *x* and *y* axes. Since variables containing uppercase letters are recommended, we will use X and Y for numbers that correspond to the *x* and *y* axes. For example, let "X=1:100;" where X is a sequence of numbers from 1 to 100, and then let "Y=2*X-20;." This means Y=2X–20. Since the value of Y is calculated for each element of the vector X, Y will be a vector with the same number of elements as X. Here, we have

```
plot(X,Y);
```

You should see a graph like the one in Fig. 1.5. *plot* is also a type of function, but it is also called a *plot* command (or instruction) because it not only performs calculations but also draws a figure on the screen. If you want to save the displayed image, you can use the "**File**" button or the file icon shown by the arrow in Fig. 1.5 to save the contents of this window in a standard image format such as TIFF, PNG, JPEG, etc. In the case of Octave, you can also save the image in the same way from the "**File**" menu of the window displaying the graph.

The range of X is from 1 to 100, and the range of Y is from –18 to 180. The graph will be drawn to fit the X and Y ranges (Fig. 1.5). This is very convenient because you don't need to give any specific instructions to draw the appropriate graph. If the line is too thin, you can make it thicker by using "plot(X,

Fig. 1.5 Y=2X-20

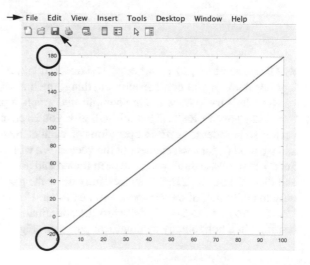

Fig. 1.6 $Y=X^2$

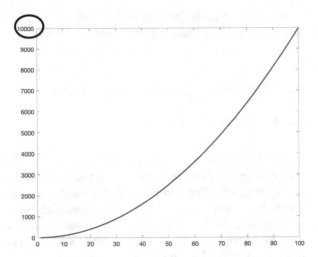

Fig. 1.7 $Y=(X-50)^2$

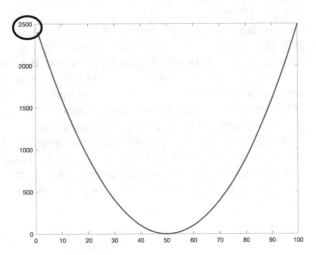

Y,'LineWidth',2);" where "'LineWidth',2" means to make the line
2 points thick (if you don't specify anything, the line will be 1 point thick).

Now, let's try to draw a more complicated graph, a parabola with $Y=X^2$, where
"Y=X^2;" or "Y=X*X;," but this will give you an error. The problem is that X is a
vector, so in order to perform operations on each element of the vector, you need to
add a period (.) for each element of the vector. We will explain this problem again in
Sect. 1.8, so please think you have been fooled and use "Y=X.^2;" or "Y=X.*X;"
and then "plot(X,Y);" to successfully draw the parabola (Fig. 1.6). In this case,
only the right half of the parabola will be drawn. However, "Y=(X-50).^2;" or
"Y=(X-50).*(X-50);" will shift the coordinates by 50 in the x-axis direction,
resulting in a horizontally symmetrical parabola (Fig. 1.7).

Fig. 1.8 Y=cos(T)

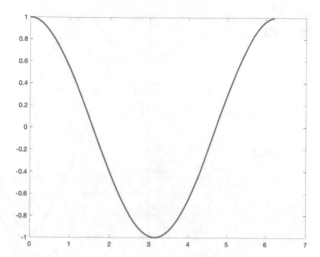

Next, consider a circle of radius 1 centered at the origin in a polar coordinate system (see Fig. 2.7 if you have forgotten polar coordinates). Let θ be the angle of a point on the circumference of the circle as seen from the origin (since we cannot use the Greek letter θ, we will use T, the initial letter of theta), and plot the *x*- and *y*-coordinates of a point on the circumference as T is varied from 0 to 2π. First, try typing "T=0:1/10:2*pi;". In this case, T will be a sequence of numbers from 0 to 2π in increments of tenths (of course, you can also use "T=0:0.1:2*pi;"). The *x*-coordinate of a point on the circumference is given by X=cosθ and the *y*-coordinate by Y=sinθ, so in MATLAB "X=cos(T); Y=sin(T);." If we give a vector T to the trigonometric function, the corresponding vectors X and Y will be calculated. "plot(T,X);" will plot X with T as the abscissa, and we will get the result as shown in Fig. 1.8. "plot(T,Y);" will plot Y with respect to T.

1.6.2 Applications of plot *Function*

plot can draw multiple graphs simultaneously. Try "plot(T,X,T,Y);." As shown in Fig. 1.9, the X values are drawn in blue and the Y values in red. More graphs can be drawn at the same time if the parentheses are arranged in the order of horizontal axis, vertical axis, horizontal axis, vertical axis, and so on.

It is also very useful to draw the graph in different colors automatically, but of course, you can change the colors yourself. "plot(T,X,'g',T,Y,'m');" will draw X in green and Y in magenta. In this case, the parentheses are in the order of horizontal axis, vertical axis, "**color**," and between the quotation marks ('') is the color information. **k**, **w**, **b**, **g**, **r**, **c**, **m**, and **y** mean black, white, blue, green, red, cyan, magenta, and yellow, respectively.

Fig. 1.9 Y=cos(T) and
Y=sin(T)

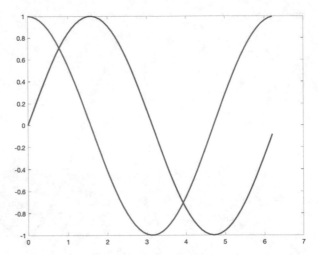

Fig. 1.10 Changing line
style

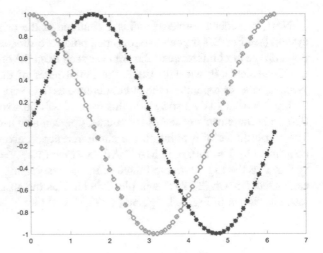

Between the quotation marks, you can add not only color information but also information about the shape of the line or point. For example, "`:`" indicates a dotted line, "`--`" indicates a dashed line, "`o`" (the letter o) indicates a circle, and "`*`" indicates an asterisk.

Now try to type "`plot(T,X,'g--o',T,Y,'m:*');.`" As shown in Fig. 1.10, X is drawn with a green dashed line and circle, and Y with a magenta dotted line and *. For more information, please search for *plot* in the MATLAB documentation (Fig. 1.1).

Now, how can we draw a circle? In that case, you can just plot X and Y without using T, in other words, "`plot(X,Y);.`" You can also add a title to the graph: "`title('circle');`" to give the graph a title, "`xlabel('X-axis');`"

Fig. 1.11 A circle

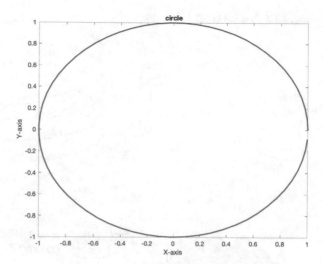

ylabel('Y-axis');" to add a description of the *x*-axis and *y*-axis, respectively (Fig. 1.11).

As you can see, *plot* is a very useful command, but you can also use *plot3* to draw 3D graphs easily. Type "plot3(X,Y,T);." The graph will be drawn with X as the *x*-coordinate, Y as the *y*-coordinate, and T as the *z* coordinate. But now let's change T a little more: "T=0:1/100:10*pi;" where θ varies from 0 to 10π in increments of 1/100, that is, a spiral going around the circumference. However, if you type "plot3(X,Y,T);", you will get an error. The workspace shows that the number of elements in T is 3142, but X and Y are still 21. Since we changed the content of T in the first place, we need to recalculate X and Y. Type "X=cos(T); Y=sin(T);" and then type "plot3(X,Y,T);."

At this point, click in the command window (the command window is active) and press the "**up**" cursor key. The command you just executed will be displayed in the command window. If you continue to press the "**up**" and "**down**" cursor keys, you can see the history of the previous commands. Then, when you see the command you want to execute again, press the Enter key to execute that command. So in this case, you can find and execute "X=cos(T); Y=sin(T);" using the cursor keys "**up**" and "**down**" and then execute "plot3(X,Y,T);" to save the trouble of typing the same command from the keyboard. In addition, "title('Circle'); xlabel ('X-axis'); ylabel('Y-axis'); zlabel('Z-axis');" will display a nice spiral with labels for the *x*, *y*, and *z* axes, as shown in Fig. 1.12.

Exercise 1.6.2 Plot the following functions using *plot* (X=0:0.01:pi).

(a) Y1=$\frac{4}{\pi}$ sin X, Y2=$\frac{4}{3\pi}$ sin 3X, Y3=$\frac{4}{5\pi}$ sin 5X, Y4=$\frac{4}{7\pi}$ sin 7X
(b) Ysum=Y1+Y2+Y3+Y4

You will find the Fourier series of a periodic function: $f(x) = \begin{cases} -1 \ (-\pi < x < 0) \\ 1 \ (0 < x < \pi) \end{cases}$

Fig. 1.12 A spiral in 3D

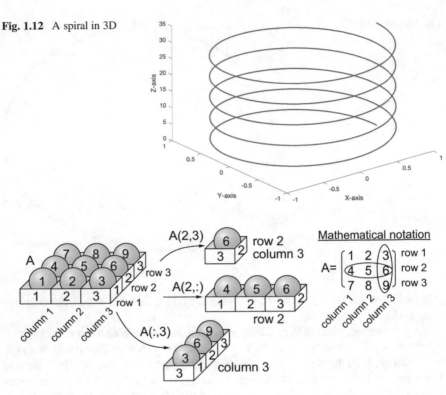

Fig. 1.13 Intuitive understanding of matrices

1.7 Matrices

1.7.1 Defining a Matrix

Now, the strength of MATLAB is that it can handle matrices freely. A matrix is a concept developed from a vector, and in the case of a two-dimensional matrix, you can think of it as a box partitioned in the vertical and horizontal directions (Fig. 1.13). The horizontal arrangement of these partitions is called rows, and the vertical arrangement is called columns. Since vectors are a type of matrix, MATLAB handles matrices in the same way as vectors. The most basic way to represent a matrix is to use brackets, "[ ]." For example, type

A= [1 2 3] ; (insert a space between 1 and 2 and between 2 and 3).

Then A will be a 1-row, 3-column (1-by-3) matrix (a vector with three elements) with 1 2 3 as its elements. For matrices with multiple rows, use a semicolon (;) to break the line.

A= [1 2 3; 4 5 6; 7 8 9];

where A is a 3-row, 3-column (3-by-3) matrix with 1 2 3 in the first row, 4 5 6 in the second row, and 7 8 9 in the third row. In Fig. 1.13, the vertical direction represents the row numbers, and the horizontal direction represents the column numbers. In mathematical notation,

$$A = \begin{pmatrix} 1 & 2 & 3 \\ 4 & 5 & 6 \\ 7 & 8 & 9 \end{pmatrix}$$

is used to represent the row number in the vertical direction and the column number in the horizontal direction (Fig. 1.13, right).

If you type "A" and press Enter, the entire contents of matrix A will be displayed, so try it. It is also possible to display only a part of the matrix.

```
A(2,3)
```

will display 6 since it represents the second row and third column of matrix A (Fig. 1.13). The first number in parentheses represents the number of rows (2), and the next number represents the number of columns (3). You can use a colon (:) here to specify a whole row or a whole column:

```
A(2,:)
```

for the entire second row, and

```
A(:,3)
```

will display the entire contents of the third column (Fig. 1.13).

It is also possible to replace some of the contents of such a matrix:

```
A(2,3)=0;
```

Type "A" and press Enter to see the result. Similarly, if you type

```
A(2,:)=0
```

all the columns in row 2 will become zero.

Now, since the matrix A has three rows and three columns,

```
A(3,3)
```

will display 9 in row 3, column 3. Now, type

```
A(3,4)
```

The fourth column will not exist and you will get the error message: "Index in position 2 exceeds array bounds (must not exceed 3)." The index is the number of the row or column in the matrix. The message warns you that the second index "4" exceeds the column number 3. But here,

```
A(:,4)=1;
```

will not produce an error message. It means to assign 1 to all rows in column 4, but in such a case, it will add a new column 4 and assign 1 to it. Type "A" and press Enter to see if it works. Conversely, if you type

```
A=A(1:2,1:2);
```

Only the 1st–2nd rows and 1st–2nd columns of matrix A will be assigned to A, resulting in a smaller matrix A.

Now, in the above example, we defined the matrix by writing all the elements of the matrix using brackets. But it is tedious to do that every time. The case where all the elements of the matrix are the same can be expressed more concisely as follows.

```
B=zeros(2,3);
```

B will be assigned to a 2-row, 3-column (2-by-3) matrix with all zero elements. *zeros* means that it will create a matrix with all elements 0. Similarly,

B=ones(3,4);

will assign 1 to a 3-by-4 matrix. *ones* means that it will create a matrix with all elements 1. The use of *ones* allows you to assign any value to the elements of B by multiplying any number you want. Type

B=pi*ones(3,4);

Then B will be a 3-by-4 matrix and all elements are π. You can also use

B=rand(4,3);

Then B will be a 4-by-3 matrix of random numbers. Type "B," press Enter, and check the elements of B. Here, we have a random number in the range of 0 to 1, but by multiplying this value by an appropriate value, we can create a matrix of random numbers in various ranges (see Sect. 2.1.2 for details).

In Sect. 1.5.1, we used *sum* to find the sum of the elements of a vector, but what happens if we apply this to a matrix? First, check the contents of matrix A, and then type "sum(A)." You will not see the sum of all the elements of the matrix, but rather

ans =

12 15 18

This is the sum of each column of matrix A. (For example, in Fig. 1.13, the first column is $1 + 4 + 7 = 12$.) Since the result displayed is a matrix with one row and three columns, and it is a vector, if we process this result with *sum* again, it should calculate the sum of all the elements of the matrix. In other words, we can do "sum (sum(A))." Alternatively, "sum(A, 'all')" will compute the sum of all the elements of matrix A at once. For more details, please search *sum* in the MATLAB documentation (Fig. 1.1).

Exercise 1.7.1a Generate the following matrices:

$$X = \begin{pmatrix} 0 & 0 & 1 & 1 \\ 0 & 0 & 1 & 1 \\ 0 & 0 & 1 & 1 \end{pmatrix} \quad Y = \begin{pmatrix} 1 & 1 & 1 & 1 \\ 1 & 1 & 1 & 1 \\ 5 & 5 & 5 & 5 \end{pmatrix}$$

Exercise 1.7.1b Calculate the sum of each column of X and Y.

Exercise 1.7.1c Calculate the sum of all the elements of X and Y.

1.7.2 The Size of a Matrix

In some cases, you may not know the size of a matrix, that is, how many rows and columns it has. In such cases, there is a way to find out the size of a matrix by using the function *size*.

If you type "size(A)" and press Enter, you will get

ans=

2 2

This means that the size of A is two rows and two columns. Similarly, if you type

size(B)

you will see that the size of B is four rows and three columns. In this example, both the number of rows and the number of columns are shown, but in some cases, it is only necessary to know how many rows or columns there are. Here,

size(A,1)

displays only the number of rows in A, and

size(A,2)

displays only the number of columns in A. In other words, in the parentheses of the size function, 1 refers to the rows and 2 refers to the columns.

Exercise 1.7.2 You can generate 3D matrices by "zeros(X,Y,Z)," "ones(X,Y,Z)," and "rand(X,Y,Z)" (here, X,Y,Z are integers larger than 0). Assign a 3D matrix of any size to M and find the size of its third dimension using *size*.

1.8 Calculating Matrices

1.8.1 Sum and Product of Matrices

Of course, we can also do operations on matrices. Let's assign a 2-by-3 matrix to A as follows.

A=[1 2 3; 4 5 6];

Here,

A+1

will display the matrix with all elements of A plus 1. Also

A*2

will display the matrix of all the elements of A multiplied by 2.

Here, assign another matrix B.

B=[1 2; 3 4; 5 6];

If you type

A+B

You will get the error message: "Matrix dimensions must agree." This is not surprising since A has two rows and three columns, and B has three rows and two columns, and the sizes of the matrices do not match, so they cannot be added together. However, if you type

A*B

You will get

ans =

22 28

49 64

This follows the mathematical definition of the product of matrices, which is the operation shown in Fig. 1.14. In other words, each row of A and each column of B is

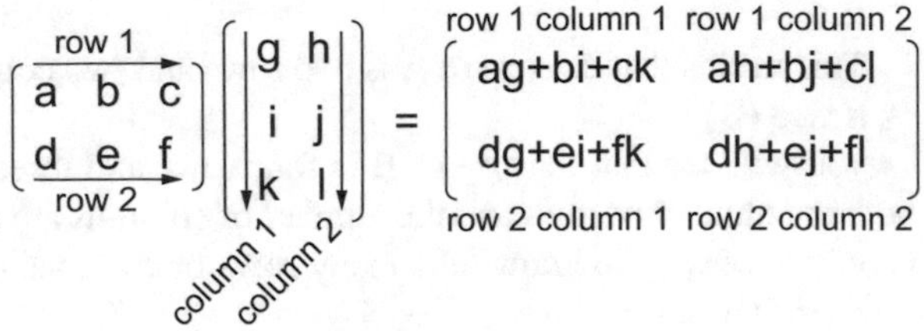

Fig. 1.14 Product of matrices

$$\begin{pmatrix} \overrightarrow{\begin{array}{ccc} a & b & c \\ \underline{d} & e & f \end{array}} \\ \text{row 2} \end{pmatrix} \begin{pmatrix} \begin{array}{cc} g & h \\ i & j \\ k & l \end{array} \end{pmatrix} = \begin{pmatrix} ag{+}bi{+}ck & ah{+}bj{+}cl \\ dg{+}ei{+}fk & dh{+}ej{+}fl \end{pmatrix}$$

row 1 column 1 row 1 column 2

row 2 column 1 row 2 column 2

considered as a vector, and the **inner product** of each vector is calculated. Thus, in a matrix product, the number of columns in the first matrix (A in this case) and the number of rows in the second matrix (B in this case) must be the same. The product of matrices according to the original definition will not appear in this book later, so if you have never learned matrices before, please do not worry about it.

On the other hand, in order to calculate the sum of matrices, the size of the two matrices, that is, the number of rows and the number of columns, must be the same. Here, A has two rows and three columns, and B has three rows and two columns, so we cannot calculate the sum as it is, but we can change A+B to

    A+B'

Now you should see the result of the calculation. Here, the single quotation mark (') means **transpose**, and it works to swap the rows and columns of the matrix (i.e., row n becomes column n and column m becomes row m). If we type

    B

you get

    1 2
    3 4
    5 6

If you transpose B as follows,

    B'

You will get

    1 3 5
    2 4 6

and it becomes a matrix of the same size as A, so we can calculate the sum of each element of A and B. Let's try.

1.8.2 Hadamard Product

The product of matrices as shown above is the original definition, but in fact, such a product of matrices does not appear again in this book. In the sum of matrices, we simply added each of the corresponding elements of the matrix together. Similarly, the product of corresponding elements, the **Hadamard product**, is more common (Fig. 1.15). Here, we have

    D=rand(2,3)

Fig. 1.15 Hadamard
product of matrices

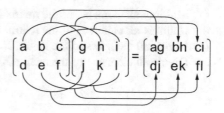

Then, for example, we have

D = 0.8147 0.1270 0.6324

   0.9058 0.9134 0.0975

Each element of the matrix is a random number, so the individual values will be different from yours, but the fact that there are two rows and three columns is the same. Also, the contents of A should be as follows (we assigned to A at the beginning of the previous section).

A = 1 2 3

   4 5 6

In other words, A and D are the same 2-by-3 matrix, but when we try to compute A*D, we get an error message:

Error using *

Incorrect dimensions for matrix multiplication. Check that the number of columns in the first matrix matches the number of rows in the second matrix. To perform elementwise multiplication, use '.*'

According to the suggestion, let's prefix the "*" with "." before and type

A.*D

You get

ans =

0.8147 0.2540 1.8971

3.6232 4.5669 0.5852

In other words, 0.8147 in the upper left corner, row 1, column 1 is 1 times 0.8147, and 0.5852 in the lower right corner, row 2, column 3 is 6 times 0.0975. As you can see, adding a period (.) calculates the product of corresponding elements, the Hadamard product (Fig. 1.15).

Similarly, in division,

A./D

calculates the value of each element of A divided by corresponding element of D. The same is true for power (^).

D.^A

You may be confused about whether to prefix or postfix the "*" and "/" operations, but you need to **prefix** before doing these operations. This will be very important later on, so please keep it in mind.

Exercise 1.8.2 Multiplying a number N to random numbers between 0 and 1 given by *rand* provides random numbers between 0 and N. Assign a 4-by-5 2D matrix with random numbers between 0 and 10 to X and Y.

(a) Multiply all elements of the matrix X by 10.
(b) Multiply all elements of the matrix X by the corresponding elements of Y.
(c) Divide all elements of the matrix X by 10.
(d) Divide all elements of the matrix X by the corresponding elements of Y.

Chapter 2
Introduction to MATLAB Programming

2.1 Scripting

2.1.1 Creating a Script

So far, we have been entering commands one by one in the command window, but there is a limit to what we can do with this method. Programming is a process of combining dozens or hundreds of commands that you have used in the past, and in order to do so, you need to create a script (program) that lists the commands (in MATLAB, the program is called a script, but in this book, it is mainly called a program).

Select "**HOME tab**" from the tabs on the upper left of the MATLAB window, click "**New Script**" on the leftmost tab or "**New**" on the third tab from the left, and then select "**Script**" (Fig. 2.1). A new area called "**Editor**" will appear above the command window, and an Editor tab labeled "**untitled**" will appear (Fig. 2.2). This is where you create your program.

In Octave, click on the new script icon in Fig. 2.3 to create a new script. The Command Window tab will switch to the Editor tab. Use the tabs at the bottom of the screen to switch between the Command Window and the Editor.

The program consists of many commands, but they are basically executed one by one from top to bottom. If you click on the white space to the right of the 1, a flashing bar will appear, and you can start typing your program here.

In the first line, please type

```
%script2_1A
```

and press Enter. Then a 2 will appear below the 1, and you can now type the second line. This "%script2_1A" has no particular meaning. In fact, the "%" symbol means to ignore the remainder of the line. You may think that it is meaningless if it is ignored, but it is useful to have an explanation when you review your program later. This kind of description is called a **comment**, and you can write a note

© The Author(s), under exclusive license to Springer Nature Singapore Pte Ltd. 2022
M. Sato, *Getting Started in Mathematical Life Sciences*, Theoretical Biology,
https://doi.org/10.1007/978-981-19-8257-6_2

Fig. 2.1 Creating a script

Fig. 2.2 Editing a script in Editor

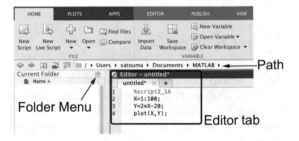

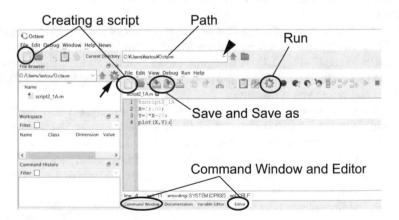

Fig. 2.3 Creating and saving a script in Octave

after the "%" to explain what kind of program it is and what kind of processing it does.

At this point, we want to save the script, but before we do that, we need to check the "**path**," which is the location where the script will be saved. In Windows and Mac, folders contain multiple nested folders and files. However, the computer is basically designed to focus on a single folder to view its contents and save files. In

Fig. 2.4 Saving a script

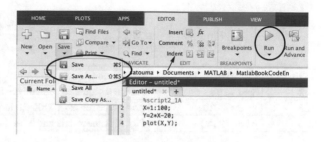

MATLAB, the arrowed boxes in Fig. 2.2 indicate paths, and in this example, the hard disk contains nested folders "Users," "satouma," "Documents," and "MATLAB," and we are currently focusing on the "MATLAB" folder.

In the Editor tab, select **"Save"** (third from the left), then **"Save As,"** and save the file as "script2_1A" (Fig. 2.4). The ".m" extension of the MATLAB script will be added and saved in the MATLAB folder as "script2_1A.m." You can then simply save the file or use the shortcut "Command+S" (on Mac) or "Ctrl+S" (on Windows). In the "Current Folder" on the left side of the window, you will see the name of the file in the "MATLAB" folder, and the name of the script you just saved will appear.

If you want to create a new folder in the "MATLAB" folder, please create a new folder from the folder menu shown in Fig. 2.2, then click "MATLAB" in the path box, and then select the newly created folder. The path to the new folder will now be specified and the files will be saved in this folder.

In Octave, click on the **"Save"** and **"Save As"** icons in Fig. 2.3 to save the script, but the installation of Octave does not create a folder for Octave. Create a new **"Octave"** folder (directory) from the icon shown by the arrow in Fig. 2.3. The box with the arrowhead indicates the path; select the newly created Octave folder and change the path.

The script used in this book can be downloaded from GitHub (https://github.com/satouma7/MatlabBookCodeEn/).

2.1.2 Creating a Script to Draw a Graph

Let's try to make a program that draws a two-dimensional graph. As in Sect. 1.6.1, when the x coordinate varies from 1 to 100, we will calculate "Y=2X–20" and plot the result. This can be written as "X=1:100;" and "Y=2*X-20;". To plot this, we can use "plot(X,Y);," which looks like "script2_1A.m" below. Note that the explanations written on the right side of each line are explanations of the process of each line, so there is no need to type them in.

script2_1A.m

1	%script2_1A	Explanation of script2_1A.m
2	X=1:100;	Assign a sequence of numbers from 1 to 100 to the vector X
3	Y=2*X-20;	Assign the result of 2×X–20 to vector Y
4	plot(X,Y);	Plot X as the *x*-axis and Y as the *y*-axis

Type this in, save it, and click "**Run**" (the green triangle icon) from the Editor tab (Fig. 2.4). In Octave, click on the Run icon in Fig. 2.3 to execute the script. You can also execute the script from the command window by typing the name of the script ("script2_1A" or "script2_1A.m" in this case) and pressing Enter. Note that the ".m" extension is automatically completed whether you write it or not, so don't worry about it.

Now, let's modify this program a little so that the value of Y becomes the square of X. This can be done by changing the third line to "Y=X.^2;." Note that we need to prefix the "^" with ".". The first line should be "%script2_1B." Save the file as "script2_1B.m" and click "**Run**."

script2_1B.m

1	%script 2_1B	
2	X=1:100;	
3	Y=X.^2;	Assign X^2 to vector Y
4	plot(X,Y);	

When executed, a graph like the one in Fig. 1.6 can be drawn. It is important to note that since the *x* coordinate is a vector with 100 elements, the *y* coordinate must also have 100 elements. In this case, we have "Y=X.^2;" and Y will be a vector of the same size as X, so no need to worry. Let's save the program as "script2_1B."

Next, let's make the *y* coordinate a random number between 0 and 100. A random number between 0 and 1 is given by the *rand* function, but to make the number of elements in the Y coordinate 100, we use "Y=rand(1,100);". 1 and 100 in parentheses means a 1-by-100 matrix, or a vector of 100 elements. Since X and Y in "plot(X,Y)" need to be vectors of the same size, we need to make Y a vector of 100 elements to match "X=1:100" (or Y can be a 100-row and 1-column matrix, "Y=rand(100,1);"). The annotations in the first line such as "%script2_1B" are omitted below. Also, unless otherwise noted, always save a new script with a name for each assignment after this. Let's call this program "script2_1C.m."

script2_1C.m

1	X=1:100;	
2	Y=rand(1,100);	Assign a random number sequence with 100 elements to vector Y
3	plot(X,Y);	

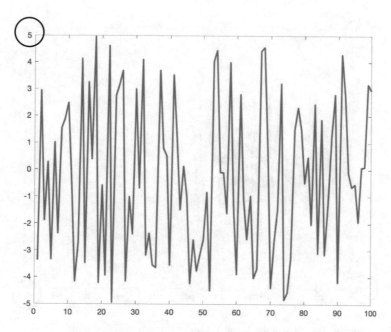

Fig. 2.5 Plotting random values of −5 to 5

When executed, the graph will look like Fig. 2.5. Now, how can we make Y a random number in the range of 0 to 10, instead of 0 to 1? We can simply multiply by 10. In other words, the second line of "script2_1C.m" should be "Y=rand (1,100)*10;." *rand* outputs a random number between 0 and 1, so if the value is 0, it will be 0 when multiplied by 10, but if the value is 1, it will be 10 when multiplied by 10. Thus, by multiplying *rand* by any number you want, you can generate random numbers of various sizes.

How can we make Y a random number from −5 to +5? To do so, we can combine multiplication and subtraction, that is, "Y=rand(1,100)*10-5;" in the second line of "script2_1C.m."

script2_1D.m: modified from script2_1C.m

2	`Y=rand(1,100)` `*10-5;`	Assign to Y a random number sequence with 100 elements in the range of −5 to 5

Using the same logic as before, if *rand* gives 0, multiply by 10 to get 0 and then subtract 5 from there to get −5. If the value is 1, multiply it by 10, subtract 5 from it, and you get +5. As you can see, although *rand* itself only generates random numbers

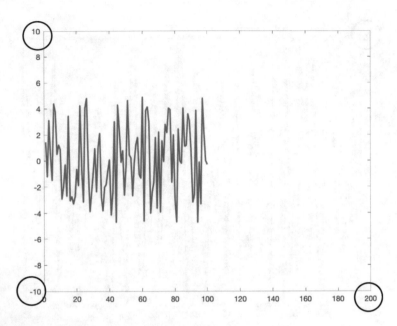

Fig. 2.6 Changing the range of the X and Y axes

between 0 and 1, you can use this value to generate random numbers in a variety of ranges.

However, the drawn graph will not look much different from Fig. 2.5. This is because MATLAB automatically adjusts the range of the graph to fit in the window, but in some cases, you may want to set the range of the graph by yourself. In that case, you can use the commands *xlim* and *ylim*. *ylim* is a command to set the range in the *y*-axis direction, and you can use it like "ylim([lower limit upper limit]);." For example, "ylim([0 10]);" will set the *y*-axis range from Y=0 to Y=10. *xlim* will set the *x*-axis range in the same way. They work on the last graph drawn. Add "xlim([0 200]); ylim([-10 10]);" to the third line of "script2_1D.m" and make it "script2_1E.m."

script2_1E.m: modified from script2_1D.m

3	plot(X,Y); xlim([0 200]); ylim([-10 10]);	Plot the x-axis range from 0 to 200 and the y-axis range from −10 to 10

When executed, the display range of the *x* and *y* axes will change as shown in Fig. 2.6.

The *rand* command used here returns a random number uniformly distributed between 0 and 1, but if you need a random number following a normal distribution (Gaussian distribution), you can use the *randn* command. It can be used in the same way as *rand*, except that the distribution will obey the Gaussian distribution (mean 0, standard deviation 1).

2.2 Circles and Spirals

2.2.1 Using Polar Coordinates to Draw Circles and Spirals

In the previous section, we graphed random numbers. Now let's try drawing circles and spirals. From the "**Home**" tab, click "**New Script**" to open a new script (Fig. 2.1).

In the case of a circle with a radius of 1 and a center at the origin, the relationship between the x and y coordinates on the circumference is $x^2 + y^2 = 1$. Now, once the range of x is determined, $y = \pm\sqrt{1 - x^2}$, but this is cumbersome because we need to consider the cases where y is positive and negative separately. In such cases, it is simpler to use polar coordinates. It is common to use θ to represent the angle of a circle in polar coordinates, but instead of θ, we will use T (Fig. 2.7). Consider the radius R=1, and let T be a vector changing from 0 to 2π, where "T=0:2*pi;." 2π is about 6.283, so the contents of T are 0, 1, 2, 3, 4, 5, 6. This only gives us the coordinates of 7 points on the circumference. To make a smooth connection on the circumference, we need more coordinates, say 100. For example, if we put ":1/100:" between 0 and 2π and write "T=0:1/100:2*pi;," T will be a vector that increases from 0 to 2π in increments of 0.01.

Once we have such a T, we can represent the points on the circumference of a circle of radius 1 using the trigonometric functions *cos* and *sin*. In other words, "X=cos(T);" and "Y=sin(T);" (Fig. 2.7). Since T is a vector, X and Y are also vectors. "cos(T)" is calculated for each element of T and assigned to vector X, and "sin(T)" is calculated for each element of T and assigned to vector Y. By plotting with the vectors X and Y prepared in this way, we can draw a circle. In other words, "plot(X,Y);." If the radius of the circle is R, then X=R cosθ and Y=R sinθ, so "X=R*cos(T); Y=R*sin(T);." Also, we need to put a specific number in R. To make a circle with a radius of 5, we just need to set "R=5;" so the program should look like "script2_2A.m" below.

Fig. 2.7 Polar coordinates

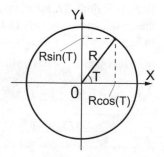

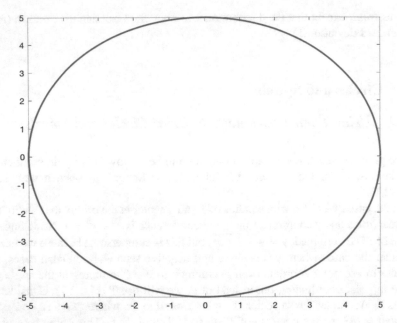

Fig. 2.8 Circle of radius 5

script2_2A.m

1	`T=0:1/100:2*pi;`	Assign a sequence of numbers in the range 0 to 2π with tick size 0.01 to T
2	`R=5;`	Assign 5 to R
3	`X=R*cos(T);Y=R*sin(T);`	Assign RcosT to X and RsinT to Y
4	`plot(X,Y);`	Plot X as the *x*-axis and Y as the *y*-axis

When executed, a circle of radius 5 is drawn as shown in Fig. 2.8.

Let's change the first line of "script2_2A.m" to "`T=0:1/100:10*pi;`," and use *plot3* to draw a three-dimensional graph. To do so, we need to change the fourth line of "script2_2A.m" to "`plot3(X,Y,T);`." In this case, T is considered to be the *z* axis, and we get the result shown in Fig. 2.9. Let's name the file "script2_2B. m."

script2_2B.m

1	`T=0:1/100:10*pi;`	
2	`R=5;`	

(continued)

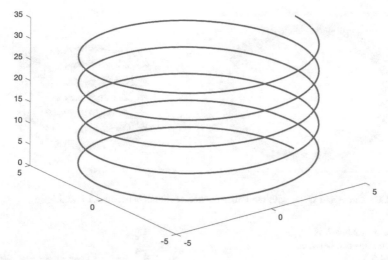

Fig. 2.9 Spiral of radius 5

3	`X=R*cos(T); Y=R*sin(T);`	
4	`plot3(X,Y,T);`	Plot X as *x*-axis, Y as *y*-axis, and T as *z*-axis

If you do "`plot(X,Y);`" and then "`plot3(X,Y,T);`," the first graph will be overwritten by the second one. The command *figure* will prepare a new window, so modify lines 4 of "script2_2B.m" as follows:

4		`figure; plot(X,Y);`
5		`figure; plot3(X,Y,T);`

This will draw two-dimensional and three-dimensional graphs in two windows, respectively. These circles may look a little distorted and oval. To make them perfect circles, add "`axis equal;`" or "`axis square;`" after line 4. Both commands specify the ratio of the lengths of the *x*, *y*, and *z* axes of the drawn graph to 1:1:1. The former sets the ratio to the unit length of the data, while the latter sets the ratio to the total length of the drawn axes.

For example, let's draw a horizontal ellipse with "`X=2*R*cos(T); Y=R*sin (T);`" in line 3. "`axis equal;`" will draw an ellipse twice as long horizontally because the ratio of the unit length in the *x* and *y* axes is 1:1, while "`axis square;`" will result in a 1:1 ratio of −10 to 10 on the horizontal axis to −5 to 5 on the vertical axis, resulting in a perfect circle.

If you run such a program repeatedly, the number of windows will increase, so close the windows by yourself.

Exercise 2.2.1a Draw a wavering spiral as shown in Fig. 2.10 ("script2_2C.m").

In "script2_2B.m," the radius was fixed at R=5, but here we need to make the radius R a vector of random numbers. Fig. 2.11 shows the relationship between T

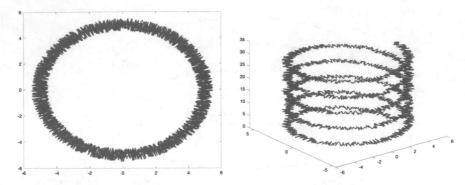

Fig. 2.10 Circle and spiral with random sequence of radius between 4.5 and 5.5

Fig. 2.11 Radius R is
random sequence between
4.5 and 5.5

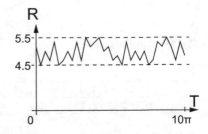

and R. *rand* gives us random numbers in the range 0–1, but how can we make the range 4.5–5.5? Yes, you can do it as 5.5-*rand* (or 4.5+*rand*, of course).

How do we calculate the number of elements in a vector? If you look at the workspace, you will see that T is 1×3142 double, where double means the precision of the number stored in T, but you don't need to think too much about it here (search the Internet for "double precision" if you are curious). 1×3142 means a matrix with 1 row and 3142 columns. T is a vector of angles on the circumference, varying from 0 to 10π. Since each angle T must have a different radius, the radius R must also be a 1-by-3142 matrix. Thus, we can set "R=5.5-rand(1,3142);" for the third row of "script2_2B.m." This looks good at first glance, but if you run it as is, you will get an error message in the command window.

This is because there is a problem with the third line, "X=R*cos(T); Y=R*sin(T);" where T is a 1-by-3142 matrix, so the result of "cos(T)" and "sin(T)" will be the same 1-by-3142 matrix as T. "R*cos(T)" and "R*sin (T)" are products of matrices. Here, we want to compute the product of each element of the matrix. As explained in the previous chapter on matrix computation, such a product is different from the original matrix product and is called the Hadamard product (Fig. 1.15). Therefore, we need to prefix the "*" with a period (.) and use ".*."

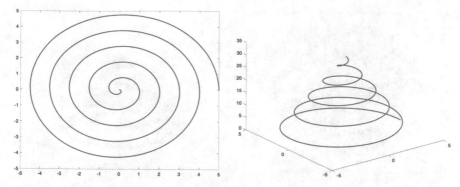

Fig. 2.12 Circle and spiral with radius changing from 5 to 0

Fig. 2.13 Radius *R* is
changing from 5 to 0

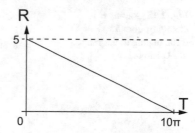

script2_2C.m

1	`T=0:1/100:10*pi;`	
2	`R=5.5-rand(1,3142);`	Assign a random number sequence in the range 4.5 to 5.5 to R
3	`X=R.*cos(T); Y=R.*sin(T);`	Assign RcosT to X and RsinT to Y
4	`figure; plot(X,Y);`	
5	`figure; plot3(X,Y,T);`	

You can also use "`R=5.5-rand(1,size(T,2));`" for the second line. The *size* command allows you to find out the number of elements in vector T without looking at the workspace. T is a matrix with 1 row and 3142 columns. Here, "`size(T,2)`" is used to find out the number of columns, not rows.

Exercise 2.2.1b Draw a spiral whose radius varies from 5 to 0 as shown in Fig. 2.12 ("script2_2D.m").

The elements of the matrix R will change with the value of T. T varies from 0 to 10π, but when T is 0, the radius is 5, and when T is 10π, the radius is 0. The image in Fig. 2.13 shows the relationship between T and R. The horizontal axis is T, and the vertical axis is R. Consider a line that passes through (T, R) = (0, 5) and (10π, 0). Since the slope is $-5/10\pi$ and the line passes through (T, R)=(0, 5), we can write "`R=-T/2/pi+5;`."

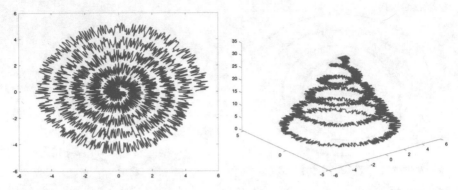

Fig. 2.14 Circle and spiral with radius changing from 5 to 0 fluctuating in the range of –0.5 to 0.5

Fig. 2.15 Radius *R* is changing from 5 to 0 fluctuating in the range of –0.5 to 0.5

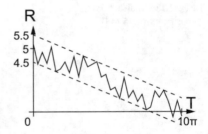

script2_2D.m: modified from "script2_2C.m"

2	R=-T/2/pi+5;	Assign a sequence of numbers decreasing from 5 to 0 to R

Exercise 2.2.1c Draw a whirlpool whose radius varies from 5 to 0 while fluctuating randomly in the range of plus or minus 0.5, as shown in Fig. 2.14 ("script2_2E. m"). The image in Fig. 2.15 shows the relationship between T and R.

script2_2E.m: modified from "script2_2C.m"

2	R=-T/2/pi+5+rand (1,3142)-0.5;	Assign a sequence of numbers decreasing from 5 to 0 while fluctuating in the range of +, −0.5 to R

2.3 *for* Statement

2.3.1 *Creating a Loop with* **for** *Statement*

There are various programming languages in the world, but the basic idea, basic functions, and commands are common. The *for* statement is a typical example of such basic commands. It is used to repeat a series of operations (loops) over and over, starting with *for* and ending with *end*. In particular, it plays a central role in the simulation of differential equations by solving them using a computer in Chap. 3 and later. Here, we will use the following example: "**Assign the result of 1+2+3+4+. . . +100 to the variable S1**." In this case, a program such as the following "script2_3A. m" can be used to add the larger numbers one by one. Read the explanations on the right side of the program, and try to figure out what it does.

script2_3A.m		
1	`S1=0;`	Assign 0 to S1 at first
2	`for I=1:100`	Increment I from 1 to 100 (I=1 at first)
3	`S1=S1+I;`	Add I to S1 and assign it to S1 again
4	`end`	Return to the second line
5	`S1`	Display the content of S1

The first line assigns 0 to S1, resetting the value to 0 since it would cause problems if S1 already contained some other value. In the second line, after the *for* statement, we have "$I=1:100$." This notation is similar to that of vectors, but here, I is not treated as a vector but simply as a scalar variable. In the third line, "$S1=S1+I;$." Since S1=0 and I=1, the result of 0+1, that is, 1, is assigned to S1.

Next, the end of the fourth line jumps back to the *for* statement in the second line, where 1 is added to the value of I, and the third line is executed with I=2. Since S1=1, the result of 1+2, that is, 3, is assigned to S1. The same process is repeated between lines 2 and 4, with each iteration adding 1 to the value of I, until finally I becomes 100.

The semicolon (;) after $S1=S1+1$ in the third line does not interfere with the operation, but without it, the result of the calculation will be displayed in the command window every time S1 is calculated. Since it would be unsightly and take extra time to process the program, we have added a semicolon to avoid displaying the calculation result. Note that there is no semicolon after `for I=1: 100` in the second line and end in the fourth line. It is a rule not to put a semicolon after a *for* statement or an *end* statement.

In this way, S1 stores the result of the calculation with values from 1 to 100 added. Finally, to display the value of S1, we use "$S1$" in line 5. Since there is no semicolon, the value of S1 will be displayed (in this case S1 should be 5050). If it works, try changing the second line of "script2_3A.m" to "`for I=1:2:100`" and run it.

Multiple instructions can be placed between this *for* statement and *end*. For example, how can we perform both of "**Assign the result of 1+2+3+4+.... to the variable S1**" and "**Assign the result of $1^2+2^2+3^2+4^2+...+100^2$ to variable S2**" at the same time? Just prepare S1 and S2 as before, and add the line "`S2=S2+I^2;`."

script2_3B.m

1	`S1=0; S2=0;`	Assign 0 to S1 and S2
2	`for I=1:100`	Increment I from 1 to 100
3	`    S1=S1+I;`	Add I to S1 and substitute S1 again
4	`    S2=S2+I^2;`	Add the square of I to S2 and substitute S2 again
5	`end`	Return to the second line
6	`S1`	Display the contents of S1
7	`S2`	Display the contents of S1

The result should be S1=5050, S2=338350.

2.3.2   *Nesting* **for** *Statements*

You can also create a loop of a *for* statement within a loop of a *for* statement. This kind of loop within a loop is called **nesting**, and the nesting of *for* statements makes it possible to perform calculations on each element of a two- or three-dimensional matrix. For example, let's say you want to compute the sum of all the elements of a 3-by-4 matrix of random numbers from 0 to 1, and the sum of the squares of all the elements. Consider the following program:

script2_3C.m

1	`A=rand(3,4);`	Assign random numbers to the 3-by-4 matrix A
2	`S1=0; S2=0;`	Assign 0 to S1 and S2
3	`for I=1:3`	Increment I from 1 to 3 (rows)
4	`  for J=1:4`	Increment J from 1 to 4 (columns)
5	`    S1=S1+A(I,J);`	Add the elements of A to S1 and assign to S1
6	`    S2=S2+A(I,J)^2;`	Add the square of the elements of A to S2 and assign to S2
7	`  end`	Return to line 4
8	`end`	Return to line 5
9	`S1`	Display the contents of S1
10	`S2`	Display the contents of S2

In the first line, we have a 3-by-4 matrix A consisting of random numbers, and in the second line, we have S1 and S2 containing 0. Lines 3–8 are nested *for* statements, but it is not difficult if you follow them line by line. Here, I represents the number of rows and J represents the number of columns. At the very beginning, I=1 is assigned in the third line, J=1 in the fourth line, and in the fifth and sixth lines, S1 and S2 are calculated using the values in the first row and first column of matrix A. In lines 5–6, S1 and S2 are calculated using the values of row 1 and column 1 of matrix A. Then, *end* in line 7 returns us to line 4. Since we are going back to the corresponding *for* statement, we are going back to line 4 instead of line 3, where J becomes 2, and lines

5–6 again compute S1 and S2 using the values of row 1 and column 2 of matrix A. Lines 4–7 loop as J changes from 1 to 4, keeping I=1 fixed. Then, from the end of line 8, we go back to the *for* statement in line 3, and the loop in lines 4–7 is executed so that J changes from 1 to 4 while I is fixed at I=2. In this way, every time I increases by 1, the loop in lines 4–7 is executed. Finally, in lines 9–10, the contents of S1 and S2 are displayed and the process ends. In this example, the nesting of the *for* statement is double, but it can be as many times as you want. The important point is to change the variable in each loop of the *for* statement, and in "script2_3C.m," I and J are used separately.

Indentations are often used to make it easier to see the structure of the *for* statement loops. For example, lines 4–7 are shifted to the right a little more than the other lines, and lines 5–6 are shifted further to the right. You can edit the indentation by using the Tab key to move it to the right, or Shift+Tab to move it to the left. In the "**Editor tab**," there is an item called "**Indent**," and you can use the icon to select multiple lines at the same time to increase or decrease the indentation (Fig. 2.4 arrows).

2.4 Converting Numbers to Integers

Next, I would like to explain how to convert a number to an integer. Integerization is the process of converting a number with a decimal point to an integer. There are several methods for rounding (rounding up, rounding down, rounding to the nearest whole number):

fix: Round down to the nearest whole number.

ceil: Round up to the nearest whole number.

round: Round off the decimal point.

For example, prepare a 3-by-3 random matrix in the command window with "A=10*rand(3)" (a single number in parentheses means a square 2D matrix with the same number of rows and columns). After checking the contents of A, type "fix (A)," "ceil(A)," "round(A)," and press Enter. You can see what these functions are doing. This is integerization.

Now for an application problem, create a 5-by-5 matrix B consisting of random integers in the following ranges in the command window:

(1) 50 to 60;

(2) −10 to +10.

There are many ways to do this, for example, "round(10*rand(5))+50" for (1), and "round(20*rand(5))-10" for (2).

You can also round to a specified number of digits, not just to the nearest whole number. For example, "round(pi)" will return a value of 3, but "round (pi,1)" will round the value to one decimal place, resulting in 3.1, where the second number represents the number of decimal places. Conversely, you can round to the left of the decimal point by specifying a negative value. For example, "round

(100*pi,-1)" will round off the first place to 310. Try this in the command window.

2.5   Characterization of Numbers

Computers are essentially only capable of handling numbers. But in reality, they handle characters such as the alphabet. In fact, each character has a corresponding code number, and the computer uses that number to handle the character. In addition, numbers such as 0 and 9 are also assigned a code number as a character. All the characters displayed on our computer screens are represented as a series of numbers inside the computer. We are usually not aware of the numerical values assigned to each character. But if you don't know this, you will have problems in programming. In MATLAB, variables are basically assigned numerical values, and all the variables in this book so far have been assigned numerical values. But you can also assign letters and text to variables using a set of **single** quotation marks. Type the following in the command window:

```
A='Good morning';
```

You can also use the command *disp* to display the contents of a variable containing characters (or even a variable containing a number):

```
disp(A);
```

You will see that the contents of A will indeed be displayed. However, this is not very helpful because the contents of A are displayed even if you simply write A and press Enter. This *disp* command is useful when you want to display a more complex text.

2.6   Mixed Number and Character Sentences

Even if a complex program produces a calculation result, it may be difficult to understand what it means if the number is just displayed. Therefore, it would be useful if you could add an explanation to the number so that the meaning of the number can be easily understood. For example, let's consider a program that displays "My favorite number is X" (where X is a random integer between 1 and 10). For example, "X=ceil(10*rand);" will assign a random integer between 1 and 10 to X. If you omit the parentheses and the number after *rand*, it will generate a random number (scalar) in the range of 0 to 1. The problem is how to connect the string "My favorite number is" and X to display them. Here, string stands for a series of characters.

Two useful commands are *num2str* and *strcat*, where *num2str* stands for "number to string," that is, to convert a number to a string. "`num2str(X)`" converts a number X to a string X.

strcat is an instruction to concatenate multiple strings into a single string. For example, if you set two strings, "A= 'MATLAB is ';" and "B='just MATLAB';", type "C=strcat(A,B);". The contents of C will be "MATLAB is just MATLAB."

So, to print "My favorite number is X," we can combine *disp*, *strcat*, and *num2str* from the previous section as follows

```
disp(strcat('My favorite number is ',num2str(X) ));
```

I think it is a little complicated and hard to understand. If you use the wrong number or position of parentheses, you will not get the result you want, so be careful.

Now, please consider the following "script2_6A.m" as an application of the previous content. Let's create a matrix M consisting of four rows and five columns of random integers from 0 to 10, and print all the elements of the matrix in the format "Row Y column X is Z." A program like the following could be considered, using a double loop of *for* statements to display the elements of matrix M one by one.

script2_6A.m	

1	`M=round(10*rand(4,5));`	Assign random integers between 0 and 10 to the 4-by-5 matrix M
2	`for I=1:4`	Increment I from 1 to 4 (rows)
3	`  for J=1:5`	Increment J from 1 to 5 (columns)
4	`    disp(strcat('Row', num2str(I), 'column', num2str(J), 'is ', num2str(M(I,J)) ));`	
		Display the value of row I and column J of matrix M
5	`  end`	Return to line 3
6	`end`	Return to line 2

In line 4, there are so many parentheses that it may be difficult to type them correctly. If you click on the right side of each parenthesis in the editor window, the parenthesis you are looking at will be underlined, and the corresponding parenthesis will also be underlined (in Octave, the parentheses will change color). This feature can be used to reduce errors.

When the program is completed, check the contents of M from the command window after execution to make sure that the results displayed are correct. It is also possible to omit *strcat*, in which case "`strcat( )`" is replaced by brackets, "`[ ]`". You can also use "`disp( ['Row', num2str(I), 'column' num2str(J), 'is ', num2str(M(I,J))] )`" in the fourth line of "script2_6A.m."

If this works well, consider the following "script2_6B.m," which is a bit more advanced. Let's create an A-by-B matrix M consisting of 0–10 random integers (where A is a random integer between 3 and 5, and B is a random integer between 4 and 7), print the size of the matrix as "In a matrix of A rows and B columns," and also print all the elements of the matrix as "Row Y column X is Z." In other words, the size of the matrix itself is randomized. For example, the program would look like the following:

script2_6B.m

#	Code	Description
1	`A=ceil(3*rand)+2;`	Assign a random integer between 3 and 5 to A
2	`B=ceil(4*rand)+3;`	Assign a random integer between 4 and 7 to B
3	`M=round(10*rand(A,B));` Assign a random integer of 0–10 to the A-by-B matrix M	
4	`disp(strcat('In a matrix of ', num2str(A),' row ',` `num2str(B),' columns'));`	
		Display the number of rows (A) and columns (B) of matrix M
5	`for I=1:A`	Increment I from 1 to A (rows)
6	`  for J=1:B`	Increment J from 1 to B (columns)
7	`    disp(strcat('Row ',num2str(I),' column ', num2str(J),` `' is ', num2str(M(I,J)) ));`	
		Display the values of row A and column B of matrix M
8	`  end`	Return to line 6
9	`end`	Return to line 5

Again, after executing this, check the contents of M from the command window and make sure that the results displayed are correct. Note that in lines 1–2, if *rand* returns exactly 0, you will get A=2 or B=3. This is unlikely to happen, but if you are concerned about it, you can use "A=round(2*rand)+3;" or "B=round(3*rand)+4;".[1]

---

[1] You can also use the command *randi*, which returns a uniform random integer. For example, "randi(3)" will give you random integers between 1 and 3, so in the first line of "script2_6B.m," you can use "A=randi(3)+2;."

2.7 *while* **Statement**

A similar statement to the *for* statement is the *while* statement, which can be used in place of the *for* statement, and in some cases may be more suitable than the *for* statement (you can skip this section, as it is less common than the *for* statement). In the above example, the range of variables I and J specified by the *for* statement was specified in advance, such as 1 to A, 1 to B, and so on. On the other hand, when the range is not known in advance, the *while* statement is useful. For example, let's assign a random sequence of integers from 0 to 10 to a matrix M with 1 row and 100 columns, check the value of element Y in column X starting from column 1, and print "X-th value is Y" as long as it is less than 10. In such cases, the *while* statement is useful. For example, a program like the following "script2_7A.m" can be considered:

script2_7A.m

1	`M=round(rand(1,100)*10);`	Assign random integers from 0 to 10 to a vector M with 100 elements
2	`I=1;`	Assign 1 to I
3	`while M(1,I)<10`	If the I-th element of M is less than 10, execute the loop up to line 6
4	`  disp(strcat(num2str(I),'-th value is ',num2str(M(1,I)) ));`	Display the I-th value of M (if I is greater than 100, an error occurs)
5	`  I=I+1;`	Add 1 to I
6	`end`	Return to line 3

The first line prepares a matrix M with 1 row and 100 columns, that is, a vector with 100 elements, and assigns random numbers from 0 to 10. In the *for* statement, "for I=1:100," I is automatically incremented by one, but in the *while* statement, I must be specified as "I=I+1;" (line 5). The third line says "while M(1,I) <10," so lines 4–5 are executed only when the I-th element of M is less than 10, that is, between 1 and 9. If it is 10, the program ends there. As in the case of the *for* statement, there is no semicolon (;) after the *while* statement and the *end* statement.

After running the program, type "M(1,I)" in the command window and check the value of M(1,I) when the *while* statement loop from line 3 stops. The value of M (1,I) should be 10.

However, there is a problem with this program, and in some cases, an error will occur on line 3. If there are 100 random numbers, at least one of them will be 10, and it is quite possible that the loop of the *while* statement will stop. If there are only 10 random numbers, the probability that none of the values in M(1,I) will be 10 is much higher. Now change the first line to "M=round(rand(1,10)*10);" and run it. M will become a sequence of 10 random numbers, instead of 100 numbers. If you do this several times, you will get the error message:

```
Index in position 2 exceeds array bounds (must not exceed 10).
Error in script2_7Awhile (line 3)
while M(1,I)<10
```

If there is no 10, the *while* statement will not stop looping even if I is larger than the number of elements in M, and the value of I will keep growing and exceed 10. Since M contains only 10 elements, an error occurs on line 3 if I is larger than 10.

To prevent such an error, we need to add another restriction to the *while* statement in line 3, that is, we need to stop the loop not only when the element of M is less than 10, but also when I is greater than 10. Let's modify lines 1 and 3 of "script2_7A.m" to make it "script2_7B.m":

script2_7B.m: modified from "script2_7A.m"

| 1 | M=round(rand(1,10) *10); | Assign random integers from 0 to 10 to a vector M with 10 elements |
| 2 | while (I<11)&(M(1, I)<10) | Execute the loop if I is less than 11 and the I-th element of M is less than 10 |

In this way, the loop after the *while* statement is executed only when I is less than 11 and the I-th element of M is less than 10, thus avoiding the above error.

The "&" in the third line will be highlighted in the editor, and if you move the mouse pointer closer, you will see a message saying, 'If any argument is a numeric scalar, consider replacing & with && for performance reasons." The process of "&" and "&&" is almost the same, but since "&&" is faster, let's just change it to "&&" (there is no problem even if you don't change it). The result is as follows:

| 3 | | while (I<11)&&(M(1,I)<10) |

The third line of "script2_7B.m" is a little complicated. If you combine the *for* statement with the *break* statement, you can perform the same operation more concisely:

script2_7C.m

1	M=round(rand(1,10)*10);	
2	for I=1:10	
3	if M(1,I)==10	
4	break	Terminate the *for* loop when M(1,I)==10
5	end	
6	disp(strcat(num2str(I),'-th value is ', num2str(M(1,I))));	
7	end	

The *break* statement is to forcibly terminate the currently executing *for* loop and cancel the process up to the end of the *for* statement loop (in this case, the *end* of line 7). The *if* statement will be explained in the next section, so please review "script2_7C.m" again later.

2.8   Rock-Paper-Scissors and *if/switch* Statements

2.8.1   *Conditional Judgment by* if *Statement*

Like the *for* statement, the *if* statement is frequently used in any programming language. The *if* statement executes an instruction according to some condition. For example, if you want to assign a random integer from 1 to 3 to the variable M, and if M is 1, assign "Rock" to the Hand variable; if M is 2, assign "Paper"; if M is 3, assign "Scissors"; and finally display the contents of Hand. The program will look like "script2_8A.m" below:

script2_8A.m

1	`M=ceil(rand*3);`	Assign a random integer between 1 and 3 to M
2	`if M==1`	If M is 1, then
3	`  Hand='Rock';`	assign 'Rock' to Hand
4	`elseif M==2`	If not, but if M is 2, then
5	`  Hand='Paper';`	assign 'Paper'to Hand
6	`else`	Otherwise,
7	`  Hand='Scissors';`	assign 'Scissors' to Hand
8	`end`	The *if* statement from line 2 ends here
9	`disp(Hand);`	Display the contents of Hand

Lines 2–3 mean that "if the content of M is 1, 'Rock' is assigned to Hand." Note that there are **two equals** between "M" and "1" in line 2. If there is **one equal** (=), it means to assign the right side to the variable in the left side, but if there are two equals (==), it means to judge if the left and right sides are equal. The *end* in line 8 corresponds to the *if* in line 2, and the presence of *end* indicates that the *if* statement ends here. Following the *if* statement, we can add further judgments using *elseif* and *else*. The *elseif* in line 4 means that if M is not 1, but M is 2, then "Paper" is assigned to Hand, while *else* in line 6 means that if M is 3, then "Scissors" is assigned to Hand. As in the case of *for* and *while* statements, there is no semicolon (;) after the *if*, *elseif*, *else*, and *end* statements. Also, since *elseif* and *else* are optional, they may or may not be present.

Next, let's consider a situation where Team 1 and Team 2 are playing 10 rounds of Rock-Paper-Scissors and assign the Rock-Paper-Scissors moves of the two teams to a 2-by-10 matrix M. In other words, the first row of M represents the moves of Team 1, the second row represents the moves of Team 2, and the 10 Rock-Paper-Scissors moves are assigned to columns 1–10, respectively. When two teams of 10 players each are playing Rock-Paper-Scissors, how can we display the list of moves as "Y-th player of Team X is Rock" for all 20 players? A program like the following "script2_8B.m" can be considered:

---

script2_8B.m

1	`M=ceil(rand(2,10)*3);`Assign random integers from 1 to 3 to the 2-by-10 matrix M	
2	`for J=1:10`	Increment J from 1 to 10 (J-th player)
3	`   for I=1:2`	Increment I from 1 to 2 (Team I)
4	`      if M(I,J)==1`	If M(I,J) is 1, then
5	`         Hand='Rock';`	assign 'Rock' to Hand
6	`      elseif M(I,J)==2`	If not, but if M(I,J) is 2, then
7	`         Hand ='Paper';`	assign 'Paper' to Hand
8	`      else`	If not,
9	`         Hand ='Scissors';`	assign 'Scissors' to Hand
10	`      end`	The *if* statement from line 4 ends here
11	`      disp(strcat(num2str(J), '-th player of Team ',num2str(I),`	
	`         ' is ', Hand));`	Display the hand of J-th player from Team I
12	`   end`	Return to line 3
13	`end`	Return to line 2

In the first line, random integers from 1 to 3 are assigned to the matrix M, which has two rows and ten columns. In the loop from line 2 to 13, the variable J represents the index of columns of matrix M, that is, the index number of players on each team (from the 1st to 10th players). In the loop from line 4 to 13, the variable I represents the index of rows of matrix M, that is, the index number of teams (Teams 1 and 2). In the *if* statement from line 4 to 10, if M(I,J) is 1, then "Rock" is assigned; if 2, then "Paper" is assigned; and if 3, then "Scissors" is assigned to the variable Hand. There are many ways to write the *if* statement in lines 4 to 10, but the example in "script2_8B.m" is not very smart. It can be written more smartly using a *switch* statement.

2.8.2   *Conditional Judgment by* switch *Statement*

The *switch* statement classifies cases between *switch* and *end*, specifying the variable to be classified immediately after the *switch* and the *case* starting from the next line, case X, case Y, and so on. For example, take a look at the following "script2_8C.m."

script2_8C.m	
1 `M=ceil(rand(2,10)*3);`	Assign random integers from 1 to 3 to a 2-by-10 matrix M
2 `for J=1:10`	Increment J from 1 to 10 (J-th player)
3 `  for I=1:2`	Increment I from 1 to 2 (Team I)
4 `    switch M(I,J)`	Classify cases according to the value of M(I,J)
5 `    case 1`	If M(I,J) is 1, then
6 `      Hand='Rock';`	assign 'Rock' to Hand
7 `    case 2`	If M(I,J) is 2, then
8 `      Hand='Paper';`	assign 'Paper' to Hand
9 `    case 3`	If M(I,J) is 3, then
10 `      Hand='Scissors';`	assign 'Scissors' to Hand
11 `    end`	The *switch* statement from line 4 ends here
12 `    disp(strcat(num2str(J), 'th player of Team ',num2str(I),`	
`      ' is ', Hand));`	Display the hand of J-th player from Team 1
13 `  end`	Return to line 3
14 `end`	Return to line 2

In line 4, we indicate that M(I,J) will be classified into cases using the *switch* statement. If the content of M(I,J) is 1, "Rock" is assigned (lines 5–6); if the content is 2, "Paper" is assigned (lines 7–8); and if the content is 3, "Scissors" is assigned (lines 9–10) to Hand. This is much easier to read and understand than the *if* statement. Note that, as in the case of the *if* statement, there is no semicolon (;) after the *switch*, *case*, and *end* statements.

Exercise 2.8.2   Complete the Rock-Paper-Scissors game

Now, it is not enough to just display the hands of the Rock-Paper-Scissors game; let's have the two teams play Rock-Paper-Scissors and see who wins. Based on "script2_8C.m," have Team 1 and Team 2 play against each other, and create a program that displays the results for ten games, such as "4th player of Team 1 is Paper, 4th player of Team 2 is Rock, and Team 1 wins for the 4th game." When you have completed the program, display the number of wins, losses, and draws for

Team 1 and Team 2, respectively, at the end. For example, "Team 1 has 5 wins, 3 losses, and 2 draws; Team 2 has 3 wins, 5 losses, and 2 draws."

As a hint, why don't you prepare a "Result" variable in the same way as the Hand variable and assign "Team 1 wins," "Team 2 wins," and "Draw" according to the results? Also, how about preparing "Team1," "Team2," and "Draws" variables and using each of them to count the number of Team1 wins, Team2 wins, draws? The following "script2_8D.m" is an example:

script2_8D.m

1	`M=ceil(rand(2,10)*3);`	Assign random integers from 1 to 3 to a 2-by-10 matrix M
2	`Team1=0; Team2=0; Draws=0;`	Team1, Team2, and Draws are the numbers of Team 1 wins, Team 2 wins, and draws
3	`for J=1:10`	Increment J from 1 to 10 (J-th player)
4	`for I=1:2`	Increment I from 1 to 2 (Team I)
5	`switch M(I,J)`	Classify cases according to the value of M (I,J)
6	`case 1`	If M(I,J) is 1, then
7	`Hand='Rock';`	assign 'Rock' to Hand
8	`case 2`	If M(I,J) is 2, then
9	`Hand='Paper';`	assign 'Paper' to Hand
10	`case 3`	If M(I,J) is 3, then
11	`Hand='Scissors';`	assign 'Scissors' to Hand
12	`end`	The *switch* statement from line 5 ends here
13	`disp(strcat(num2str(J), 'th player of Team ',num2str(I), ' is ', Hand));`	Display the hand of J-th player from Team I
14	`end`	Return to line 4
15	`switch M(1,J)`	Classify cases according to the value of M (1,J)
16	`case 1`	If the J-th player of Team 1 is 'Rock'
17	`if M(2,J)==2`	If the J-th player of Team 2 is 'Paper'
18	`Result='Team 2 wins'; Team2=Team2+1;`	Team2 wins and adds 1 to Team2
19	`elseif M(2,J)==3`	If not, but if the J-th player of Team 2 is 'Scissors'
20	`Result ='Team 1 wins'; Team1=Team1+1;`	Team1 wins and add 1 to Team1
21	`else`	If not (if the J-th player of Team 2 is 'Rock')
22	`Result ='Draw'; Draws=Draws +1;`	Draw game and add 1 to Draws
23	`end`	The *if* statement from line 17 ends here

(continued)

24	`case 2`	If the J-th player of Team 1 is 'Paper'
25	`if M(2,J)==1`	If the J-th player of Team 2 is 'Rock'
26	`Result ='Team 1 wins';` `Team1=Team1+1;`	Team1 wins and add 1 to Team1
27	`elseif M(2,J)==3`	If not, but if the J-th player of Team 1 is 'Scissors'
28	`Result ='Team 2 wins';` `Team2=Team2+1;`	Team2 wins and adds 1 to Team2
29	`else`	If not (if the J-th player of Team 2 is 'Paper')
30	`Result ='Draw'; Draws = Draws` `+1;`	Draw game and add 1 to Draws
31	`end`	The *if* statement from line 25 ends here
32	`case 3`	If the J-th player of Team 1 is 'Scissors'
33	`if M(2,J)==1`	If the J-th player of Team 2 is 'Rock'
34	`Result ='Team 2 wins';` `Team2=Team2+1;`	Team2 wins and add 1 to Team2
35	`elseif M(2,J)==2`	If not, but if the J-th player of Team2 is 'Paper'
36	`Result ='Team 1 wins';` `Team1=Team1+1;`	Team1 wins and add 1 to Team1
37	`else`	If not (if the J-th player of Team 2 is 'Scissors')
38	`Result ='Draw'; Draws=Draws` `+1;`	Draw game and add 1 to Draws
39	`end`	The *if* statement from line 33 ends here
40	`end`	The *switch* statement from line 15 ends here
41	`disp(strcat(Result, ' in the ',` `num2str(J), '-th game'));`	Display the result of the J-th game
42	`end`	Return to line 3 and increase J
43	`disp(strcat('Team1 has ',num2str` `(Team1),' wins, ',`	
	`num2str(Team2),' losses, and ',` `num2str(Draws),' draws'));`	Display the number of wins, losses and draws of Team 1
44	`disp(strcat('Team2 has ',num2str` `(Team2),' wins, ',`	
	`num2str(Team1),' losses, and ',` `num2str(Draws),' draws'));`	Display the number of wins, losses and draws of Team 2

2.9   Functions

2.9.1   *Functions for Judging Rock-Paper-Scissors*

In "script2_8D.m," lines 15–41 are used for judging the Rock-Paper-Scissors game, and this alone takes up 27 lines. In this program, only one Rock-Paper-Scissors judgment is made, but in a more complex program, multiple Rock-Paper-Scissors judgments may be made over and over again. In such a case, writing 27 lines of Rock-Paper-Scissors judgments each time is tedious and makes the program difficult to read. If you define a frequently used process as a "new function" and give it a name, you can then call that function to perform the exact same process with only one instruction. In this example, the function "Judge" is defined as "Judge.m" below, which judges the winner given two numbers (i.e., Rock-Paper-Scissors hands for two players):

Judge.m

1	`function [R]=Judge(A, B)`	Define *Judge* function that receives A and B and outputs R
2	`switch A`	Classify cases according to the value of A
3	`case 1`	If player A is 'Rock'
4	`  if B==2`	If player B is 'Paper'
5	`    R=2;`	assign 2 to R (B wins)
6	`  elseif B==3`	If not, but if player B is 'Scissors'
7	`    R=1;`	assign 1 to R (A wins)
8	`  else`	If not (if player B is 'Rock')
9	`    R=0;`	assign 0 to R (draw)
10	`  end`	The *if* statement from line 4 ends here
11	`case 2`	If player A is 'Paper'
12	`  if B==1`	If player B is 'Rock'
13	`    R=2;`	assign 2 to R (B wins)
14	`  elseif B==3`	If not, but if player B is 'Scissors'
15	`    R=1;`	assign 1 to R (A wins)
16	`  else`	If not (if player B is 'Paper')
17	`    R=0;`	assign 0 to R (draw)
18	`  end`	The *if* statement from line 12 ends here
19	`case 3`	If player A is 'Scissors'
20	`  if B==1`	If player B is 'Rock'
21	`    R=2;`	assign 2 to R (B wins)
22	`  elseif B==2`	If not, but if player B is 'Paper'
23	`    R=1;`	assign 1 to R (A wins)
24	`  else`	If not (if player B is 'Scissors')
25	`    R=0;`	assign 0 to R (draw)

(continued)

| 26 | `end` | The *if* statement from line 20 ends here |
| 27 | `end` | The *switch* statement from line 2 ends here |

Enter the above code and save it as "Judge.m." "`function [R]=Judge(A, B)`" in line 1 declares a new function *Judge*, and the name of the function must match the name of the saved file "Judge" (without the ".m" extension). The *function* is to define a function, but what are "`[R]`" and "`(A,B)`"? The *Judge* function is a function that determines the winner of a Rock-Paper-Scissors game between two players. In this case, A and B represent a Rock-Paper-Scissors hand each, and the integer is either 1 (Rock), 2 (Paper), or 3 (Scissors). R is assigned the winner of the game, 1 (A wins), 2 (B wins), or 3 (draw). The reason why R is enclosed in brackets, "[ ]," is because that is how it is supposed to be written. Also, there is no need to put a semicolon (;) after a *function* or *end* statement.

In the second line, the contents of A are classified into three cases by "switch A": 1 (Rock; lines 3–10), 2 (Paper; lines 11–18), and 3 (Scissors; lines 19–26). Then, for each case, *if, elseif, else* is used to determine the winner of the Rock-Paper-Scissors game when the value of B is 1 (Rock), 2 (Paper), or 3 (Scissors), and the result is assigned to R.

Since this "Judge.m" is a function, it is not useful by itself; we need to create a program that calls the *Judge* function below. Please save this file as "script2_9A.m."

script2_9A.m

1	`M=ceil(rand(2,10)*3);`	Assign random integers from 1 to 3 to a 2-by-10 matrix M
2	`Team1=0; Team2=0; Draws=0;`	Team1, Team2, and Draws are the numbers of Team1 wins, Team2 wins, and draws
3	`for J=1:10`	Increment J from 1 to 10 (J-th player)
4	`  for I=1:2`	Increment I from 1 to 2 (Team I)
5	`    switch M(I,J)`	Classify cases according to the value of M (I,J)
6	`    case 1`	If M(I,J) is 1, then
7	`      Hand='Rock';`	assign 'Rock' to Hand
8	`    case 2`	If M(I,J) is 2, then
9	`      Hand='Paper';`	assign 'Paper' to Hand
10	`    case 3`	If M(I,J) is 3, then
11	`      Hand='Scissors';`	assign 'Scissors' to Hand
12	`    end`	The *switch* statement from line 5 ends here
13	`    disp(strcat(num2str(J), 'th player of Team ',num2str(I), ' is ', Hand));`	Display the hand of J-th player from Team I
14	`  end`	Return to line 4

(continued)

15	`Decision=Judge(M(1,J),M(2,` `J));`	Assign the result of *Judge* between J-th player of Team1 and Team2 to *Decision*
16	`switch Decision`	Classify cases according to the value of Decision
17	`case 1`	If Decision is 1 (Team 1 wins)
18	`Result='Team1 wins';` `Team1=Team1+1;`	Team1 wins and add 1 to Team1
19	`case 2`	If Decision is 2 (Team 2 wins)
20	`Result ='Team2 wins';` `Team2=Team2+1;`	Team2 wins and add 1 to Team2
21	`case 0`	If Decision is 0 (draw)
22	`Result ='Draw';Draws=Draws+1;`	Draw game and add 1 to Draw
23	`end`	The *switch* statement from line 16 ends here
24	`disp(strcat(Result,' in the ',` `num2str(J),'th game'));`	Display the result of the J-th game
25	`end`	Return to line 3
26	`disp(strcat('Team1 has ',num2str` `(Team1),' wins, ',num2str` `(Team2),`	
	`' losses, and ', num2str` `(Draws),' draws'));`	Display the number of wins, losses, and draws of Team1
27	`disp(strcat('Team2 has ',num2str` `(Team2),' wins, ',num2str` `(Team1),`	
	`' losses, and ', num2str` `(Draws),' draws'));`	Display the number of wins, losses, and draws of Team2

Notice that in line 15, "`Decision=Judge(M(1,J),M(2,J));`" `M(1,J)` and `M(2,J)` represent the Rock-Paper-Scissors hands of Team 1 and Team 2, respectively, and the result is assigned to the Decision variable. Earlier, in the explanation of the *Judge* function, I explained that A and B are the hands of Rock-Paper-Scissors and R is the result of the game. A, B, and R are just temporary variable names inside the "Judge.m" function, and any variable name can be used when calling the *Judge* function. In line 15, "`Decision=Judge(M(1,J),M (2, J);`" "Judge.m" will be executed with `M(1,J)` assigned to A and `M(2,J)` assigned to B, and the result in R will be assigned to Decision. This may be a little complicated, but you can see that running "script2_9A.m" will give you the same result as "script2_8D.m."

2.9.2   Placing a Function in a Script

In the above example, the *Judge* function is in a separate file called "Judge.m." The fact that it is a separate file has the advantage that it can be called from other programs, but it may be disadvantageous in terms of program execution speed, since it calls the "Judge.m" file stored on the PC hard drive when the function is used. In such a case, you can incorporate the *Judge* function into the main program. Please see "script2_9B.m" below. The process is the same as "script2_9A.m" and "Judge.m."

script2_9B.m: modified from "script2_9A.m"

15	Decision=Judge2(M(1,J),M(2,J));	
28		blank line
29	function [R]=Judge2(A,B)	Define *Judge2* function that receives A and B and outputs R
30	switch A	Classify cases according to the value of A
31	case 1	If player A is 'Rock'
32	if B==2	If player B is 'Paper'
33	R=2;	assign 2 to R (B wins)
34	elseif B==3	If not, but if player B is 'Scissors'
35	R=1;	assign 1 to R (A wins)
36	else	If not (if player B is 'Rock')
37	R=0;	assign 0 to R (draw)
38	end	The *if* statement from line 32 ends here
39	case 2	If player A is 'Paper'
40	if B==1	If player B is 'Rock'
41	R=1;	assign 1 to R (A wins)
42	elseif B==3	If not, but if player B is 'Scissors'
43	R=2;	assign 2 to R (B wins)
44	else	If not (if player B is 'Paper')
45	R=0;	assign 0 to R (draw)
46	end	The *if* statement from line 40 ends here
47	case 3	If player A is 'Scissors'
48	if B==1	If player B is 'Rock'
49	R=2;	assign 2 to R (B wins)
50	elseif B==2	If not, but if player B is 'Paper'
51	R=1;	assign 1 to R (A wins)
52	else	If not (if player B is 'Scissors')
53	R=0;	assign 0 to R (draw)
54	end	The *if* statement from line 48 ends here
55	end	The *switch* statement from line 30 ends here
56	end	The *Judge2* function from line 29 ends here

Line 28 is a new blank line to make it easier to read. Basically, the rest of "script2_9B.m" copies almost the same contents as "Judge.m." However, to avoid confusion with the already saved *Judge* function, the name of the function is changed to *Judge2* (lines 15, 29). Lines 29–56 are the definition of *Judge2* function. The result is the same as "script2_9A.m."

2.10   Handling 2D Images

2.10.1   *Image Display with* imagesc

When displaying the results of a calculation based on a mathematical model, for example, if you want to plot the time variation of a protein concentration in a single cell, you can use the *plot* instruction described in Sects 1.6 and 2.1. However, when considering the temporal variation of the protein concentration in each cell in a multicellular system, it is necessary to consider the spatial spread. In this case, the *plot* instruction is insufficient. The *imagesc* instruction, on the other hand, is very useful because it represents a sheet of cells in a two-dimensional plane and visualizes the gene expression levels in color.

If X is a two-dimensional matrix and "imagesc(X)" is used, a graph reflecting the contents of X will be displayed. For example, try "X=100*rand(100);" in the command window. In this case, X is a matrix with 100 rows and 100 columns, and each element has a random value from 0 to 100.

Next, try "imagesc(X);." This will visualize the contents of the matrix X (Fig. 2.16). Here, blue represents smaller values, and the yellow represents larger values. To add a color bar, type "colorbar;." A color bar will now appear on the right side of the graph, showing the correspondence between colors and values (Fig. 2.17).

If you do not add special instructions, it will automatically show small values in blue and large values in yellow. There are many variations of such a value-color

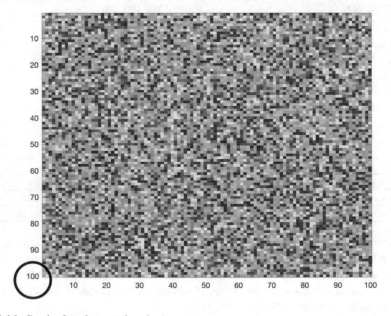

Fig. 2.16 Graph of random numbers by *imagesc*

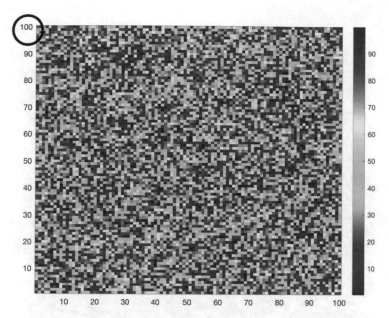

Fig. 2.17 Y-axis corrected, *jet* colormap

correspondence (**colormap**), and you can specify them as you like. By default, a colormap named *parula* is used, but you can change it using the command *colormap*. For example, a colormap that is blue for small values, light blue-green-yellow for larger values, and red for the largest values is commonly used and is named *jet*. Try typing "colormap(jet);" in the command window, and you will see that the colors of the graph and color bar displayed by *imagesc* have changed (Fig. 2.17). You can also use "colormap(gray);" to get a black-and-white grayscale colormap. For other colormaps, search for "colormap" in the MATLAB documentation (Fig. 1.1).

Here, the horizontal and vertical axes are marked from 10 to 100 in increments of 10, but the vertical axis is larger at the bottom. If you want it to be displayed upside down, type "axis xy;." Then, you will see that the direction of the vertical axis becomes normal (Fig. 2.17). It is very useful to be able to give a series of commands like this and have it redraw the graph each time (*imagesc* must be run first). If you put all these instructions together as "imagesc(X); axis xy; colorbar;" you can get the desired graph in one line.

It is also possible to display a similar graph in three dimensions: try "surface (X);." Click on the rotation arrow icon at the top of the window (Fig. 2.18 arrow), and drag the graph to rotate it to the desired angle (Fig. 2.19).

However, at this point, the two-dimensional graph that was just displayed by "imagesc(X)" has been overwritten and disappeared. Let's use the *figure* command to display the two graphs at the same time before the *imagesc* and *surface* instructions:

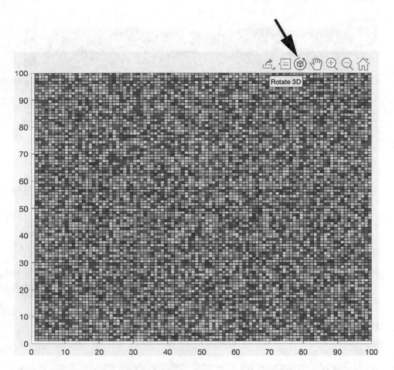

Fig. 2.18 Graph of random numbers by *surface*

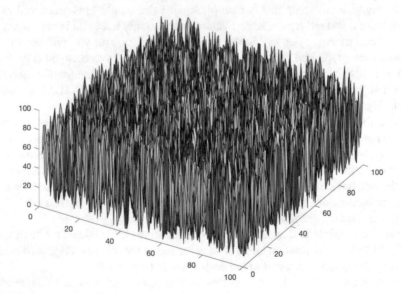

Fig. 2.19 Rotating the 3D graph

```
figure; imagesc(X); axis xy; colorbar;
figure; surface(X); colorbar;
```

The graph will be drawn in a new window.

Now, we have displayed a matrix with random values, but we want to display something more meaningful. Type the following "script2_10A.m" and run it.

script2_10A.m

1	Z=zeros(100,100);	Assign 0 to a 100-by-100 matrix Z
2	for X=1:100	Increment X from 1 to 100
3	for Y=1:100	Increment Y from 1 to 100
4	Z(Y,X)=((X-50)/10)^3+((Y-50)/4)^2;	Assign the result to row Y and column X of Z
5	end	Return to line 3
6	end	Return to line 2
7	figure; imagesc(Z); axis xy; colorbar;	Display Z contents in 2D in a new window
8	figure; surface(Z); colorbar;	Display Z contents in 3D in new window

You will see a two-dimensional graph with *imagesc* (Fig. 2.20) and a three-dimensional graph with *surface* (Fig. 2.21). Here, the fourth line represents $Z = \left(\frac{X-50}{10}\right)^3 + \left(\frac{Y-50}{4}\right)^2$, where the *for* statement changes X from 1 to 100 and Y from

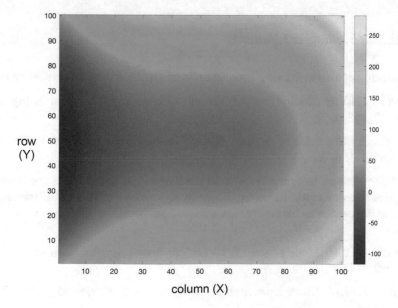

Fig. 2.20 2D graph by *imagesc*

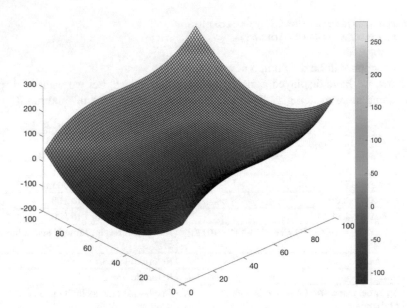

Fig. 2.21 3D graph by *surface*

1 to 100, and calculates the value of the Z element at each X and Y coordinate. It is important to note that Z(Y,X) is used instead of Z(X,Y), because Z is a two-dimensional matrix, but inside the parentheses is **row** and **column** in row-column order. Try changing the formula in line 4 to see how the result changes.

2.10.2 Creating 2D Patterns

In order to get used to drawing two-dimensional graphs, try the following exercises.

Exercise 2.10.1a Modify "script2_10A.m" to draw a graph shown in Fig. 2.22 ("script2_10B.m").

Exercise 2.10.1b Modify "script2_10A.m" to draw a graph shown in Fig. 2.23 ("script2_10C.m").

The conditional judgment "==" after the *if* statement was to determine if the left and right sides are equal. However, inequality signs such as ">" and "<" can also be used. In the case of an inequality sign with an equal sign, the equal sign is followed such as ">=" or "<=".

Exercise 2.10.1c Modify "script2_10A.m" to draw a graph shown in Fig. 2.24 ("script2_10D.m").

Exercise 2.10.1d Create a pattern like the one in Fig. 2.25 ("script2_10E1.m").

Exercise 2.10.1e Repeat the pattern in Fig. 2.25 10 times in the horizontal direction as shown in Fig. 2.26 ("script2_10E2.m").

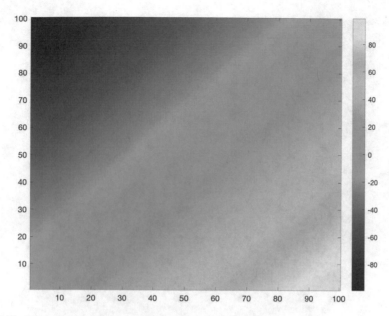

Fig. 2.22   2D pattern1

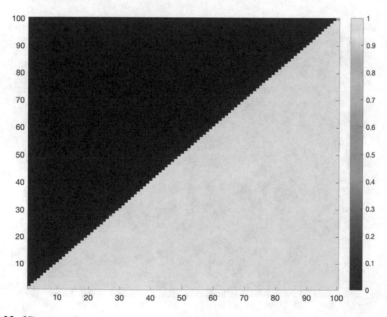

Fig. 2.23   2D pattern 2

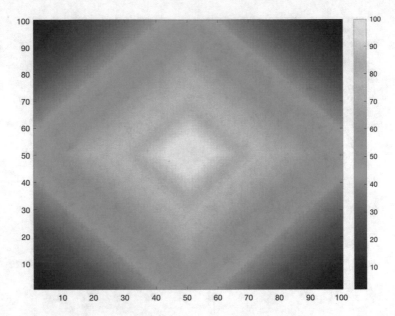

Fig. 2.24 2D pattern 3

Fig. 2.25 2D pattern 4

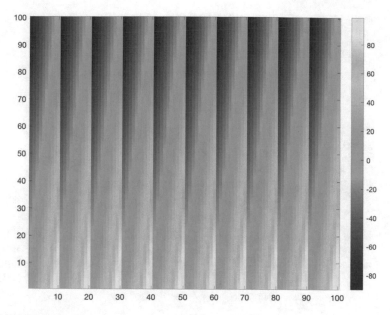

Fig. 2.26 2D pattern 5

The instruction *repmat* is useful for this kind of operation. "repmat(original matrix, number of repetitions in the vertical direction, number of repetitions in the horizontal direction);" can be used to generate a pattern that repeats the original matrix in the vertical and horizontal directions.

Exercise 2.10.1f Create a pattern like the one in Fig. 2.27 ("script2_10E3.m").

	script2_10B.m	
1	Z=zeros(100,100);	Assign 0 to a 100-by-100 matrix Z
2	for X=1:100	Increment X from 1 to 100
3	for Y=1:100	Increment Y from 1 to 100
4	Z(Y,X)=X-Y;	Assign the value X-Y to row Y and column X of Z
5	end	Return to line 3
6	end	Return to line 2
7	figure;imagesc(Z);axis xy; colorbar;	Display Z contents in 2D in a new window

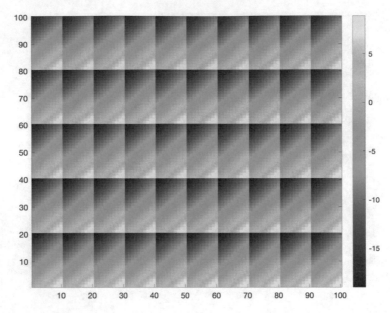

Fig. 2.27 2D pattern 6

script2_10C.m

1	`Z=zeros(100,100);`	
2	`for X=1:100`	
3	`  for Y=1:100`	
4	`  if X>=Y`	If X>=Y then (this line can be "if X>Y")
5	`  Z(Y,X)=1;`	assign 1 to row Y and column X of Z
6	`  end`	The *if* statement in line 4 ends here
7	`  end`	
8	`end`	
9	`figure;imagesc(Z);axis xy;` `colorbar;`	

script2_10D.m

1	`Z=zeros(100,100);`	
2	`for X=1:50`	Increment X from 1 to 50 (upper left area)
3	`  for Y=1:50`	Increment Y from 1 to 50
4	`  Z(Y,X)=X+Y;`	assign the value of X+Y to row Y column X of Z
5	`  end`	Return to line 3
6	`end`	Return to line 2

(continued)

7	`for X=51:100`	Increment X from 51 to 100 (upper right area)
8	`  for Y=1:50`	Increment Y from 1 to 50
9	`  Z(Y,X)=Y-X+101;`	assign the value Y−X+101 to row Y column X of Z
10	`  end`	Return to line 8
11	`end`	Return to line 7
12	`for X=1:50`	Increment X from 1 to 50 (lower left area)
13	`  for Y=51:100`	Increment Y from 51 to 100
14	`  Z(Y,X)=X-Y+101;`	assign the value X−Y+101 to row Y column X of Z
15	`  end`	Return to line 13
16	`end`	Return to line 12
17	`for X=51:100`	Increment X from 51 to 100 (lower right area)
18	`  for Y=51:100`	Increment Y from 51 to 100
19	`  Z(Y,X)=202-(X+Y);`	Assign the value 202–(X+Y) to row Y column X of Z
20	`  end`	Return to line 18
21	`end`	Return to line 17
22	`imagesc(Z);axis xy; colorbar;`	

script2_10E1.m

1	`Z=zeros(100,100);`	
2	`for X=1:10`	Increment X from 1 to 10
3	`  for Y=1:100`	Increment Y from 1 to 100
4	`  Z(Y,X)=10*X-Y;`	Assign the value of 10*X-Y to row Y column X of Z
5	`  end`	Return to line 3
6	`end`	Return to line 2
7	`figure;imagesc(Z);axis xy; colorbar;`	

`script2_10E2.m`: modified from 'script2_10E1.m'

7	`Z=repmat(Z(1:100,1: 10),1,10);`	Assign to Z the Z contents of column 1 through 10, repeated 10 times in the column direction
8	`figure;imagesc(Z); axis xy;colorbar;`	

`script2_10E3.m`: modified from 'script2_10E2.m'

3	`for Y=1:20`	Increment Y from 1 to 20
4	`Z(Y,X)=X-Y;`	Assign X-Y values to row Y and column X of Z

(continued)

7	Z=repmat(Z(1:20,1: 10),5,10);	Repeat the Z contents of rows 1–20 and columns 1–10,
		5 times in row direction and 10 times in column direction

It is important to note that it will not work if the size of the first pattern you generate is 100-by-10 in the vertical and horizontal axes. In this case, a pattern of size 20-by-10 should be created instead, and then the pattern is repeated by *repmat*.

2.11 Animation

2.11.1 Animation of plot

So far, we have drawn a static graph, but it is often necessary to animate such a pattern as it changes over time. First, let's consider how to animate a *plot* graph.

The concept of time comes into play here, and we can add the dimension of time to the matrix. For example, let's assume that X and Y are the *x* and *y* axes, and T is the time axis. Type in the command window: "X=1:100; T=1; Y=sin(4*pi/ 100*X+2*pi/100*T);." Then "plot(X, Y);" to draw the sine curve at time T=1 (Fig. 2.28). Now let "T=10; Y=sin(4*pi/100*X+2*pi/100*T);" and "plot(X, Y);" to draw the sine curve at T=10 (Fig. 2.29). You can see

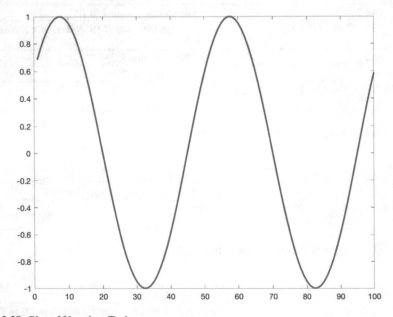

Fig. 2.28 Plot of Y at time T=1

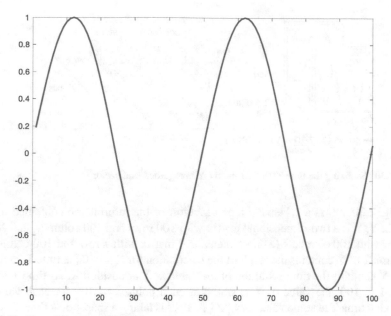

Fig. 2.29 Plot of Y at time T=10

that the sine curve is shifted along the horizontal axis as the time moves forward, so try different values of T.

Next, let's write a script to calculate the matrix Y such that the sine curve changes according to the change in T using the *for* statement. Enter the "script2_11A.m" below:

script2_11A.m

1	`Y=zeros(100,100);`	Assign 0 to a 100-by-100 matrix Y
2	`X=1:100;`	Assign a sequence of numbers from 1 to 100 to vector X
3	`for T=1:100`	Calculate the value of Y by increasing T from 1 to 100
4	`  Y(:,T)=sin(4*pi/100*X+2*pi/100*T);`	Compute the value of vector Y(:,T) at time T
5	`end`	Return to line 3
6	`for T=1:100`	Display animation by incrementing T from 1 to 100
7	`  plot(X,Y(:,T));`	plot X as horizontal axis and Y as vertical axis at time T
8	`  pause(0.1);`	Wait 0.1 seconds
9	`end`	Return to line 6

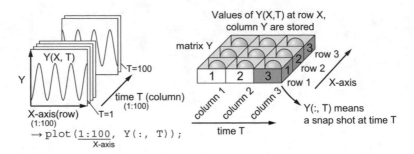

Fig. 2.30 By fixing the time T, the 2D matrix Y is regarded as a vector

Fig. 2.30 shows a schematic representation of the information contained in the matrix Y. Y is a two-dimensional matrix with 100 rows and 100 columns, and X is a vector with 100 elements (a one-dimensional matrix with 1 row and 100 columns). Column T of Y contains the result of the calculation for X=1:100 at time T, and row X of Y contains the time evolution of the value of Y at position X. As time T varies from 1 to 100, the value of Y is stored at each point X=1:100 for each value of T. Fixing time T at some value, say "Y(:,T)," is taking a snapshot at time T. Y is a two-dimensional matrix, but by specifying the time T as "Y(:,T)," it will be treated as a one-dimensional matrix (vector).

In the *for* statement loop of lines 3–5, the value of T is varied from 1 to 100, and the value of Y is calculated at each time T. In the fourth line, the calculation at time T is performed as "Y(:,T)." In MATLAB, the index number of the matrix row and column must be an integer greater than or equal to 1. Similarly, in line 2, we have "X=1:100;" so the vector X will contain values from 1 to 100. Then, in line 4, we calculate "Y(:,T)" for all the elements of vector X at time T.

Lines 6–9 are once again a *for* statement loop, where the value of T is varied from 1 to 100 to display the animation on the computer screen. If you use "plot(X, Y);" you will get an error because X is a vector and Y is a matrix with 100 rows and 100 columns, while if you use "Y(:,T)", it will be a vector with 100 elements, the same as X. Note that the size (number of elements) of the vectors must also match.

"pause(0.1)" means to wait for approximately 0.1 second. Without it, the graph at each time T will be rewritten immediately and will not animate. You can use 0.01 or 0.001 seconds instead of 0.1 seconds. Waiting a little before drawing the next time's graph will result in an animation that updates the result gradually. You can see how the sine curve moves to the left as shown in Figs. 2.28 and 2.29.

2.11.2 *Animation of* imagesc

In the previous section, we animated the *plot*, but we can do exactly the same thing for *imagesc*. In "script2_11B.m," let's set "Z=zeros(100,100,100);" to prepare a 3D matrix Z consisting of X, Y, and T coordinates. It's a little more

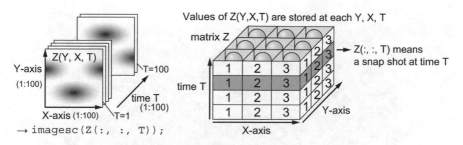

Fig. 2.31 By fixing the time T, the 3D matrix Z is regarded as a 2D matrix

complicated than "script2_11A.m." Let's calculate "Z(Y,X,T)= sin
(pi/100*X+2*pi/100*T)+2*cos(2*pi/100*Y+3*pi/100*T);" for
each X, Y, T and animate Z. This calculation is done in line 5.

Since *imagesc* is an instruction to visualize the values stored in a 2D matrix on the
plane, "imagesc(Z);" will not work because Z is a 3D matrix as shown in
Fig. 2.31. In the previous *plot*, we obtained a 1D vector by fixing the time T to the
2D matrix Y. Similarly, a 3D matrix consisting of X, Y, and T coordinates can
become a 2D matrix by fixing the time T (which means a snapshot at time T). So here
we use "imagesc(Z(:,:,T));." Here, the colon (:) means all rows and columns. It looks like the following "script2_11B.m."

script2_11B.m

1	`Z=zeros(100,100,100);`	Assign 0 to a 100-by-100-by-100 3D matrix Z
2	`for T=1:100`	Calculate the value of Z by incrementing T from 1 to 100
3	`  for X=1:100`	Increment X from 1 to 100
4	`  for Y=1:100`	Increment Y from 1 to 100
5	`  Z(Y,X,T)=sin(pi/100*X +2*pi/100*T)`	
	`+2*cos(2*pi/100*Y+3*pi/ 100*T);`	Calculate the value of Z(Y,X,T) at time T
6	`  end`	Return to line 4
7	`  end`	Return to line 3
8	`end`	Return to line 2
9	`for T=1:100`	Display animation by incrementing T from 1 to 100
10	`  imagesc(Z(:,:,T));axis xy; colorbar;`	Display the contents of the 2D matrix Z(:,:,T) at time T
11	`  pause(0.01);`	Wait 0.01 seconds
12	`end`	Return to line 9

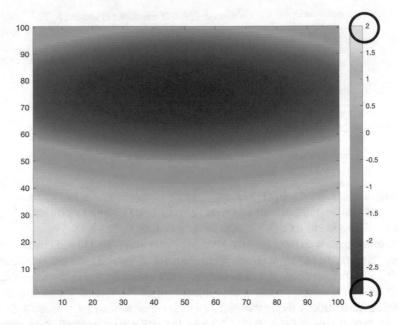

Fig. 2.32 2D animation by *imagesc*

Note that Z is a three-dimensional matrix, so the load on your computer will be greater than before. Depending on your computer, or if you are using Octave, this may take a very long time. If this is the case, try changing the value of 100 to 50 in lines 1–5 and 9.

In some cases, the execution speed may be too slow and you may want to stop the program in the middle. In that case, click on the MATLAB command window to activate it, and press the Ctrl and C keys on Windows or Command and period (.) keys on Mac at the same time to stop it.

When executed, the image will move toward the lower left corner as shown in Fig. 2.32. But at the same time, the color bar on the right side of the window will change, and the values for each color will not be consistent. This is because *imagesc* automatically changes the range of colors displayed according to the content of the displayed image. Fig. 2.32 shows the image at T=50. The color bar has a maximum value of 2 and a minimum value of –3. On the other hand, the values of *sin* and *cos* take values in the range of –1 to 1, so from "script2_11B.m," line 5, $\sin\left(\frac{\pi}{100}X + \frac{2\pi}{100}T\right) + 2\cos\left(\frac{2\pi}{100}Y + \frac{3\pi}{100}T\right)$, Z should take a value between –3 and 3. Therefore, we want to fix the maximum value of the color bar to 3 and the minimum value to –3.

You can specify a range of values for the 2D matrix to be displayed with *imagesc*. In this case, "imagesc(Z(:,:,T),[min max]);" can be used to set the minimum and maximum values for the Z display. If the value is greater than the maximum value, it will be treated as the maximum value, and if the value is less than

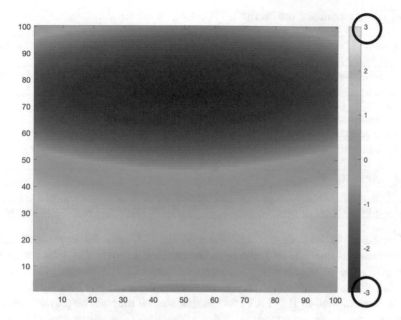

Fig. 2.33 Adjusting the value range from -3 to 3

the minimum value, it will be treated as the minimum value. Now, modify the 10th line of "script2_11B.m" as follows to make it look like "script2_11C.m":

script2_11C.m: modified from "script2_11B.m"

10	imagesc(Z(:,:,T),[-3 3]);axis xy;colorbar;

As you can see in Fig. 2.33, the animation is now displayed with the maximum value of the color bar fixed at 3 and the minimum value at −3. Try changing these values to see how the displayed image changes.

In "script2_1E.m," we used *xlim* and *ylim* to set the display range of the *plot* command. These commands can also be used for *imagesc*. For example, type "ylim ([51 80]);" in the command window, and you will see that it changes the display range of the image in the vertical axis direction.

2.11.3 Saving a Movie File

You can also save this movie as a movie file by modifying "script2_11D.m" below. Unfortunately, Octave does not implement the instructions related to saving this movie, so you will need to get MATLAB.

script2_11D.m: modified from "script2_11C.m"

9	`mov=VideoWriter ('script2_11D','MPEG-4');`	Create an object *mov* to save the movie file
10	`open(mov);`	Open the movie file based on the contents of *mov*
11	`for T=1:100`	
12	`imagesc(Z(:,:,T),[-3 3]);axis xy;colorbar;`	
13	`writeVideo(mov,getframe (gcf));`	Write one frame of the movie based on the contents of *mov*
14	`pause(0.01);`	
15	`end`	
16	`close(mov);`	Close the movie file specified by *mov*

Compared to "script2_11C.m," lines 9, 10, 13, and 16 have been added. *VideoWriter* in line 9 is an instruction for creating a movie file, which may be a little confusing because it uses the concept of "**object**." The "script2_11D" following *VideoWriter* is a file name, so you can use any file name you like. By default, the file is saved in Motion JPEG AVI format, but since it may not be playable on some devices, we have specified a video file with the extension ".mp4" as "MPEG-4."Then, create a *VideoWriter* object named *mov* with this information (think of it as a variable with advanced information). We will then use this *mov* to open and close the file, and to actually create the movie. It would be too much to explain the format of saving the movie, so for now, let's just follow the above instructions.

Line 10, "open(mov)," opens the video file and prepares to create a movie. Lines 11–15 loop to display the video frame by frame while increasing the time T by 1. In line 13, *writeVideo* writes the image that was just drawn. *gcf* stands for get current figure, which indicates the current window drawn, and *getframe* means to get the contents of the specified window as a frame of the movie. So, it means that the contents of the 2D graph drawn by *imagesc* in line 12 will be added to the movie as a frame.

After the loop in lines 11–15 is completed, "close(mov)" in line 16 closes the movie file that has been written so far and completes the process. The video file "script2_11D.mp4" should have been created in the MATLAB folder (or Octave folder), so try playing it. You can save the video by following the same procedure as in lines 9, 10, 13, and 16. So, you don't need to understand the exact meaning of each command.

Note 1. About Debugging
I hope you have understood the basic concepts of programming from the previous explanations. Basically, you should be able to do quite a lot of advanced things just by using the functions and instructions that we have

(continued)

covered so far, but in reality, when you try to create and run your own programs, you will probably find that they do not work as you expect. Even if you have mastered the commands and concepts necessary for programming, it is actually not enough to create a program that is error-free. Especially when you are a beginner in programming, you will probably be troubled by problems called **bug**.

A bug is a mistake or defect in a program. The process of removing these bugs and making the program problem-free is called **debugging**. In the past, computers were made by combining vacuum tubes, and bugs actually caused the computer to malfunction, hence the term "bug." It would have been simpler if we had only caught the bugs, but finding bugs in programs is not an easy task. In some cases, a lot of time and effort is required to debug, and a certain amount of experience is needed to debug efficiently. However, MATLAB has a lot of functions to support debugging. In later notes, I will explain some of the common types of bugs and give a brief explanation of how to find and fix them.

2.12 Gradient Rings

2.12.1 Drawing a Gradient Ring

Now that we have learned the basic programming with MATLAB, even if you understand the meaning of each command, you still need some experience to actually create the program you want. Now let's try a more difficult task. You can skip this section if you want.

"Draw a ring with an inner radius of 30 and an outer radius of 50, as shown in Fig. 2.34, and change the color along the circumference (the color value is 0–100)." If possible, I would like you to try this without any hints, but if it seems difficult, I will give you a few hints.

If you want to draw a circular diagram or graph, you can use polar coordinates, but the problem is how to bring x and y coordinates into polar coordinates. First of all, consider the centers X=50 and Y=50 of a 100-by-100 Cartesian coordinate system as the origin of the polar coordinate system, with radius R and angle θ (Fig. 2.35). Then we can color in the range of R from 30 to 50, and the color should be changed depending on θ. If the code for the color is 100 times $\theta/2\pi$, then the color will be 0 when θ=0 and 100 when θ=2π.

Once R and θ are determined, we can use trigonometric functions to find the x and y coordinates of the point. Let V be the X coordinate and W the Y coordinate viewed from the origin of the polar coordinate system, then V=Rcosθ and W=Rsinθ. You could use a *for* statement to change θ from 0 to 2π to find V and W, but then the values of V, W, X, and Y would not be integers.

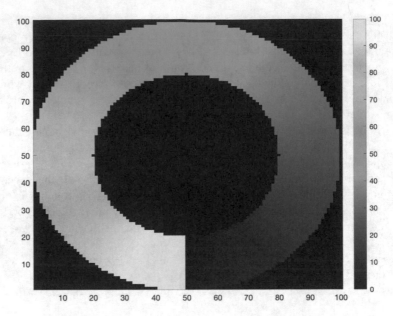

Fig. 2.34   A gradation ring

Fig. 2.35   Polar coordinates

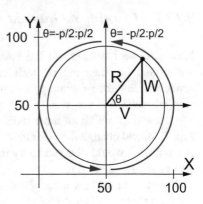

On the other hand, since we will be writing the color values into a 100-by-100 2D matrix, it is desirable that the X and Y coordinates be integers, so it is more convenient to use a *for* statement to change the X coordinate (V) and Y coordinate (W) by 1 rather than changing the value of θ.

Specifically, we will use double *for* statements to vary the X and Y from 1 to 100. For each value of X and Y, we can calculate V=X–50 and W=Y–50. Then, from Fig. 2.35, $R = \sqrt{V^2 + W^2}$, that is, "R=sqrt(V^2+W^2)." To assign a value to Z only when 30<R<50, we can use the *if* statement such as "if (R>30)&(R<50)." However, since "&&" is recommended in this case because it is faster than "&," let's say "if (R>30)&&(R<50)."

Now all we need to do is to find θ. Since tanθ=W/V, we can use the inverse function of tan, *atan*, to get θ=atan(W/V). With this in mind, try the following "script2_12A.m" code. Note that θ is rewritten as Theta.

script2_12A.m

1	`Z=zeros(100,100);`	Assign 0 to a 100-by-100 matrix Z
2	`for X=1:100`	Increment X from 1 to 100
3	`  for Y=1:100`	Increment Y from 1 to 100
4	`V=X-50;W=Y-50;`	Calculate V and W from X and Y
5	`R=sqrt(V*V+W*W);`	Calculate the distance R (radius) from the origin
6	`Theta=atan(W/V);`	Calculate θ (Theta) from V and W
7	`if (R>30)&&(R<50)`	If R is between 30 and 50,
8	`Z(Y,X)=Theta*100/2/pi;`	Calculate color value from Theta and assign to Z (Y,X)
9	`  end`	The *if* statement from line 7 ends here
10	`  end`	Back to line 3
11	`end`	Back to line 2
12	`imagesc(Z);axis xy;` `colorbar;`	

However, this will result in a pattern like the one in Fig. 2.36. There are two main problems with this code. First, we would like the Z value to vary from 0 to 100, but

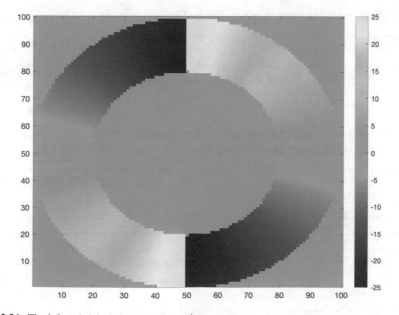

Fig. 2.36 The left and right halves are discontinuous

in fact, it ranges from –25 to +25. Check the value of the scale bar in Fig. 2.36. Also, instead of one stripe per circle, there are two borders at the bottom and top of the circle.

Look at Fig. 2.35 to see why this is the case. Consider the values of θ determined by *atan* when turning counterclockwise around the circumference of the circle from below and above (red and blue arrows). V>0 on the right side of the figure and V<0 on the left. However, for W, W>0 on the upper side and W<0 on the lower side, so the value of W/V is positive on the upper right and lower left and negative on the lower right and upper left. Since the value of θ is obtained from W/V, if we move rightward on the circumference from bottom to top (red arrow), θ will change from –π/2 to π/2. But if we move leftward from top to bottom (blue arrow), θ will likewise change from –π/2 to π/2. Therefore, we cannot distinguish the right and left halves of the circumference. We also have a problem with θ taking a negative value.

To solve these problems, we need to add an appropriate value to θ calculated by *atan*, so that θ changes from 0 to π in the right half of the circumference, and from π to 2π in the left half.

To do this, let's classify the cases into V>0 and V<0. Furthermore, when V=0, the value of W/V goes to infinity, so we need to classify it into three cases, V>0, V=0 and V<0. In the case of V>0, we need to only change the range of the *for* statement, so we don't need to use *if* statements. In the case of V>0, we need to set "for X=51:100," and we need to add π/2 to convert the range of θ (Theta) from –π/2 to π/2 to 0–π. Also, let's fix the range of values displayed by *imagesc* to 0–100 as "imagesc(Z,[0 100]);." The following "script2_12B.m" is a modified version of the previous code:

script2_12B.m : modified from "script2_12A.m"

2	for X=51:100	Increment X from 51 to 100 (right half)
6	Theta=atan(W/V)+pi/2;	Add pi/2 to atan(W/V) to get θ (Theta)
12	imagesc(Z,[0 100]);axis xy; colorbar;	

This seems to work, with the right half of the values ranging from 0 to 50 (Fig. 2.37). For V=0 (X=50), insert the following after line 11 of "script2_12B.m" and run it:

12	X=50;	Set X to 50 (for V=0)
13	for Y=1:100	
14	V=X-50;W=Y-50;	
15	R=sqrt(V*V+W*W);	
16	Theta=atan(W/V)+pi/2;	
17	if (R>30)&&(R<50)	
18	Z(Y,X)=Theta*100/2/pi;	
19	end	
20	end	
21	imagesc(Z,[0 100]);axis xy;colorbar;	

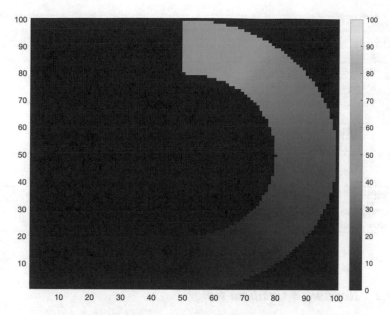

Fig. 2.37 The right half of the ring

When V=0, the value of W/V goes to infinity. However, "atan(W/V)" seems to be calculated without any problem. For example, when "atan(W/V)" is calculated in the command window with "W=10; V=0;," the result is 1.5708, or π/2. If you do the same calculation with "W=-10;," you will get −1.5708, or −π/2. Usually, division by zero is forbidden. However, since *atan* can be calculated without any problem, there is no need to separate the cases even when V=0.

Finally, if V<0, we need to use for "X=1:49" and add 3π/2 to convert the range of θ from -π/2–π/2 to π–2π. The result is shown in "script2_12C.m." Check that the result looks like Fig. 2.34.

script2_12C.m

1	Z=zeros(100,100);	
2	for X=50:100	Increment X from 50 to 100 (right half)
3	for Y=1:100	
4	V=X-50;W=Y-50;	
5	R=sqrt(V*V+W*W);	
6	Theta=atan(W/V)+pi/2;	Add pi/2 to atan(W/V) to have θ (Theta)
7	if (R>30)&&(R<50)	
8	Z(Y,X)=Theta*100/2/pi;	
9	end	
10	end	

(continued)

11	`end`	
12	`for X=1:49`	Increment X from 1 to 49 (left half)
13	`  for Y=1:100`	
14	`  V=X-50;W=Y-50;`	
15	`  R=sqrt(V*V+W*W);`	
16	`  Theta=atan(W/V)+3*pi/2;`	Add 3pi/2 to atan(W/V) to have θ (Theta)
17	`  if (R>30)&&(R<50)`	
18	`  Z(Y,X)=Theta*100/2/pi;`	
19	`  end`	
20	`  end`	
21	`end`	
22	`imagesc(Z,[0 100]);axis xy;` `colorbar;`	

2.12.2 Rotating a Gradient Ring

If the time is T, and when T changes from 1 to 100, the value of $T \times 2\pi/100$ increases to 2π. If we add this to θ, we can make the value of θ change at each time (note that θ here represents the angle and T the time). However, since θ was originally supposed to take values between 0 and 2π, this alone would cause the value to exceed 2π. In that case, the value of Z would exceed 100, which is not good since the color code is supposed to be in the range of 0–100. So, if θ exceeds 2π, we can subtract an appropriate value from θ. In Fig. 2.35, when θ exceeds 2π, we have made more than one rotation around the circumference of the circle, so we can think of it as turning back one rotation, so we subtract 2π. Here, we can use the *if* statement as follows:

```
if Theta>2*pi
  Theta=Theta-2*pi;
end
```

Exercise 2.12.1a Modify "script2_12C.m" so that the value of Theta changes with time and the gradient ring rotates once ("script2_12D.m").

script2_12D.m

1	`Z=zeros(100,100,100);`	
2	`for T=1:100`	
3	`  for X=50:100`	
4	`  for Y=1:100`	
5	`  V=X-50;W=Y-50;`	
6	`  R=sqrt(V*V+W*W);`	
7	`  Theta=atan(W/V)+2*pi/100*T+pi/2;`	Calculate Theta
8	`  if Theta>2*pi`	If Theta is greater than 2π,
9	`  Theta=Theta-2*pi;`	subtract 2π from Theta
10	`  end`	The *if* statement in line 8 ends here
11	`  if (R>30)&&(R<50)`	
12	`  Z(Y,X,T)=Theta*100/2/pi;`	
13	`  end`	
14	`  end`	
15	`  end`	
16	`for X=1:49`	
17	`  for Y=1:100`	
18	`  V=X-50;W=Y-50;`	
19	`  R=sqrt(V*V+W*W);`	
20	`  Theta=atan(W/V)+2*pi/100*T+3*pi/2;`	Calculate Theta
21	`  if Theta>2*pi`	If Theta is greater than 2π,
22	`  Theta=Theta-2*pi;`	subtract 2π from Theta
23	`  end`	The *if* statement in line 21 ends here
24	`  if (R>30)&&(R<50)`	
25	`  Z(Y,X,T)=Theta*100/2/pi;`	
26	`  end`	
27	`  end`	
28	`  end`	
29	`end`	
30	`mov=VideoWriter('script2_12D','MPEG-4');`	Prepare *mov* for saving video
31	`open(mov);`	Open a file for saving video
32	`for T=1:100`	
33	`  imagesc(Z(:,:,T),[0 100]);axis xy; colorbar;`	
34	`  writeVideo(mov, getframe(gcf));`	Write a video frame
35	`  pause(0.001);`	
36	`end`	
37	`close(mov);`	Close the video file

In the above example, "script2_12D.m," the ring in Fig. 2.34 rotates clockwise. In lines 7 and 20, "2*pi/100*T" is added to Theta. In lines 8–10 and 21–23, 2π is subtracted when Theta exceeds 2π. In line 33, "imagesc(Z(:,:,T));" is used to display a movie frame, because Z is now a 3D matrix that includes the time axis. *imagesc* only accepts 2D matrices, but by specifying the time as T, it is assumed to be 2D (see Fig. 2.31). Lines 30, 31, 34, and 37 can be omitted since they are for saving the video.

If you look closely at the code, you will see that the loop for X=50:100 (lines 3–15), which calculates the right half of the circumference, and the loop for X=1:49 (lines 16–28), which calculates the left half of the circumference, are almost identical. The only difference is that a larger constant π is added to Theta in the latter ($\pi/2$ is added in line 7, while $3\pi/2$ is added in line 20). It can be greatly simplified by adding (or not adding) π depending on the value of X. Of course, you can also use *if* statements, but let's try the following method to be a bit smarter.

Set "X=60;" in the command window. Then, if you type " (X>50)," the result will be 1. Conversely, " (X<50)" gives 0. Here, X>50 and X<50 are just inequalities. They do not imply any operation. In this case, however, X>50 is correct, but X<50 is wrong. In other words, such an inequality returns "1" if it is correct, and "0" if it is not. If you want it to determine whether they are the same or not instead of inequalities, use " (X==60)." A simple "=" means to assign, but to determine if they are the same or not, we use double equals, "==," to compare the two. We saw this in Sect. 2.8.1, when we used the *if* statement.

The value is "1" if it is correct, and "0" if it is not. This value can be used to simplify the code, that is, "+pi*(X<50)" will add π if X<50; otherwise, nothing will be added.

Exercise 2.12.1b Simplify "script2_12D.m" using the technique explained above ("script2_12E.m").

script2_12E.m

1	`Z=zeros(100,100,100);`	
2	`for T=1:100`	
3	`  for X=1:100`	
4	`  for Y=1:100`	
5	`  V=X-50;W=Y-50;`	
6	`  R=sqrt(V*V+W*W);`	
7	`  Theta=atan(W/V)+2*pi/100*T+pi/2+pi*(X&lt;50);`	Calculate Theta (if X<50 (left half), add more π)
8	`  if Theta>2*pi`	
9	`    Theta=Theta-2*pi;`	
10	`  end`	
11	`  if (R>30)&&(R<50)`	

(continued)

12	`Z(Y,X,T)=Theta*100/2/pi;`	
13	`end`	
14	`end`	
15	`end`	
16	`end`	
17	`for T=1:100`	
18	`imagesc(Z(:,:,T),[0 100]);axis xy;` `colorbar;`	
19	`pause(0.001);`	
20	`end`	

Line 7 is very important, it adds an extra π to Theta depending on the value of X by "`+pi*(X&lt;50);`".

Note 2. Common Bugs #1: Typos

Not only in MATLAB, but also in other programming languages, there are many inorganic symbols such as `:` `;` `,` `'` `(` `)` `{` `}` `[` `]`, each of which has a different meaning. Even if they look similar, they are completely different things to the computer, so don't make any mistakes. Variables are named using a combination of letters and numbers, but the upper- and lowercase letters are clearly distinguished. The names of functions and commands are often all lowercase, but sometimes they contain uppercase letters, as in *VideoWriter*. You should not misspell uppercase and lowercase.

This kind of typo bug is relatively easy to find. For example, in "script2_1A.m," if you execute "`plot(X,Y);`" as "`plot(x,Y);`" in line 4, you will get the following error message:

```
Unrecognized function or variable 'x'.
Error in script2_1A (line 4)
plot(x,Y);
```

It warns you that the variable "x" is not recognized and there is a problem in line 4. In this case, we can quickly find what the problem is and fix it based on the error message.

On the other hand, if "`X=1:100;`" is replaced with "`X=1;100;`" in line 2, you will have a blank graph and no error message. "`X=1:100;`" assigns a sequence of numbers from 1 to 100 to vector X, whereas "`X=1;100;`" just assigns 1 to scalar X. Even when the result is different from what you intended, it is difficult to find out where to fix it if there is no error message.

In this case, the graph is clearly wrong, so you can look at the values of X and Y in the workspace. Normally, both should be "1×100 double," that is, a 1-by-100 matrix. But if both X and Y are scalars, you can guess that the problem is in these variables. You can also type X or Y in the command window and check the contents of these variables.

(continued)

Note 2. Common BugsBug #1: Typos (continued)

Also, if there are many parentheses, as in the fourth line of "script2_6A.m," it is difficult to notice if there is a typing error. If there is a problem with the parentheses, you will usually get an error message, so you can focus on that line. If you click a little to the right of each parenthesis, an underline will appear on the parenthesis and its counterpart.

Chapter 3
Simulating Time Variations in Life Phenomena

In this chapter, we would like to learn how to simulate the temporal changes of life phenomena by utilizing the basics of programming we have learned so far. There are various methods of simulations, but here I will explain the method using differential equations, which can be the most basic method. You may have learned about differential and integral calculus in high school, but have you ever heard of differential equations? In short, differential equations are equations that include derivatives, but it may not be clear to you how solving differential equations is related to simulation. In fact, the content of physics taught in high school is mostly differential equations themselves.

According to Newton's equation of motion, when the velocity of motion of an object is v, acceleration is a, and time is t, the time derivative of v is a, so the relation $\frac{dv}{dt} = a$ holds. When the force on the object is constant, a is a constant (if the force is F and the mass of the object is m, then $F = ma$), and integrating both sides with t, we get $v = at + v_0$ (v_0 is the initial velocity at $t = 0$). If the position of the object is x, the time derivative of x is v, so the relation $\frac{dx}{dt} = v$ holds, and integrating this at t gives $x = \frac{1}{2}at^2 + v_0 t + x_0$ (x_0 is the initial position at $t = 0$). $\frac{dv}{dt} = a$ and $\frac{dx}{dt} = v$ are exactly differential equations, and $x = \frac{1}{2}at^2 + v_0 t + x_0$ is their solution. If you know the values of a, v_0, and x_0, you can immediately calculate the position and velocity of the object, no matter what value you apply to t. In other words, you can completely simulate the motion of the object.

Newton devised his own method of calculating mechanics, called differential and integral calculus, and used differential equations to construct Newtonian mechanics. However, the method of simulating phenomena using differential equations can be applied not only to mechanics but also to a wide range of other fields such as electromagnetism, thermodynamics, and even biological phenomena.

In the above example, the solution to the differential equation was obtained by hand calculation using paper and pencil. Such a solution obtained only by transforming the equation is called an **exact solution** of a differential equation. In such cases, you can use a calculator to calculate the position and velocity of an object

© The Author(s), under exclusive license to Springer Nature Singapore Pte Ltd. 2022
M. Sato, *Getting Started in Mathematical Life Sciences*, Theoretical Biology,
https://doi.org/10.1007/978-981-19-8257-6_3

as much as you want. However, as you will see later, the number of cases where such exact solutions can be obtained is very limited, and in most cases, you need to use a computer to solve the differential equation approximately. The solution obtained by using a computer is called a **numerical solution**, and unlike an exact solution, it contains errors due to calculation. In many cases, it is very difficult to obtain the exact solution by manual calculation, so it is important to calculate the numerical solution by programming and to perform computer simulations.

In this chapter, we will start from the molecular level phenomena such as protein synthesis and degradation to the cellular level phenomena such as proliferation and differentiation of hematopoietic stem cells to the individual level phenomena such as regulation of glucose metabolism by insulin to the population level phenomena such as infection of infectious diseases and ecology.

3.1 Synthesis and Degradation of Proteins

3.1.1 Euler Method for Solving Differential Equations

I mentioned that it is difficult to find exact solutions by hand, so we use computers to solve differential equations and find numerical solutions. But how can we use computers to handle differential equations? The original definition of differential requires the concept of **limit** as described below. But is it possible to handle this limit with a computer? Unfortunately, this is not possible. So, we have to cheat a bit and use approximate methods to calculate the numerical solution. This approximation can cause big problems, but let's start with an elementary problem that doesn't require us to think about it.

First, let's consider the simplest problem: a cell in which a protein E is synthesized and degraded. The interior of the cell is homogeneous, and we do not consider the spatial extent of the cell. Since the concentration of protein E is a function of time t, it can be expressed as $E(t)$. When we abbreviate it to E, how can we describe $\frac{dE}{dt}$, the change in E over time? If the production rate of E per unit time is p and the degradation rate coefficient is k, then it can be expressed as follows:

$$\frac{dE}{dt} = p - kE \tag{3.1.1}$$

The unit time can be 1 s, 1 min, or 1 h. The left-hand side $\frac{dE}{dt}$ represents the change in E per unit time, and the right-hand side $p - kE$ represents all the reactions that contribute to the increase or decrease in E. "**If the right-hand side contains all the chemical reactions related to E, we can completely simulate the change of E with time.**" Now, the production rate p on the right side is easy to understand. But note that the degradation rate is kE. Since E is produced whether protein E is present or not, the production rate is simply p. However, for degradation, the amount of E cannot decrease any further in the absence of E. Also, if the degradation factor k is

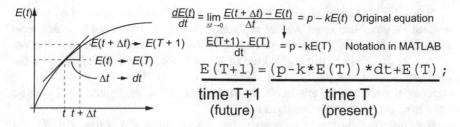

Fig. 3.1 Solving differential equations using the Euler method

0.1, it means that 10% of all E is degraded every unit of time. So even if k is constant, the actual amount of degradation depends on (and is proportional to) the amount of E. That is why the rate of degradation of E is k times E. For more details on mathematical models of reaction rates of proteins and other chemical substances, please refer to other books, for example [1].

When writing mathematical expressions, there is a rule that variables such as E and t and constants such as p and k should be written in italics, and this rule is followed in this book. Variables and constants are written in italics in mathematical expressions and in regular font style in other cases when translation of equations to MATLAB scripts is discussed. The final MATLAB scripts are written in Courier font style.

Now, as we learned in high school mathematics, the left-hand side $\frac{dE}{dt}$ is defined as $\lim_{\Delta t \to 0} \frac{E(t+\Delta t) - E(t)}{\Delta t}$, which is the slope of $E(t)$ at time t (Fig. 3.1). The tricky part is $\lim_{\Delta t \to 0}$, where we need to consider that Δt is taking the limit to zero. If we simply set $\Delta t = 0$, the denominator will be zero, which means division by zero. Here, we think of Δt as a small value that is not zero but can be considered as close to zero as possible. However, computers can only handle a finite number of fixed values, so in this case, for example, $\Delta t = 0.1$ or 0.01, or some other appropriate small value, and then proceed with the calculation. You may wonder if such a lax approach is acceptable. In fact, depending on the type of differential equation, even a small value of Δt may result in completely strange calculations. In some cases, even if you make Δt smaller, the problem may not be solved. But let's just go on with the discussion.

If we consider a small value of dt instead of Δt, Equation (3.1.1) can be written as $\frac{E(t+dt) - E(t)}{dt} = p - kE(t)$, ignoring $\lim_{\Delta t \to 0}$. Multiplying both sides by dt and transforming further, we get $E(t + dt) = (p - kE(t))dt + E(t)$ where $E(t + dt)$ and $E(t)$ are the concentrations of E at time $t + dt$ and t. Mathematically, $t + dt$ and t are real numbers and continuous values, but in the MATLAB program, they represent the time axis of a matrix or vector. As you can see in Figs. 2.30 and 2.31, the time must also be a positive integer. If the time represented by a positive integer is T, and the microtime dt is 1, then t is replaced by T and dt by 1, and we can write $E(T + 1) = (p - kE(T))\mathbf{dt} + E(T)$.

Here, T + 1, or an increase of T by 1, means that time has advanced by a small amount of time, but since we are replacing dt with 1, time should really have advanced only by dt. This problem will be explained in detail in the next section. Also, note that there is still a dt on the right-hand side, which is multiplied by the entire right-hand side of Eq. (3.1.1), that is, $(p - kE(T))$. Note that all of the differential equations in this book are in the form of dt multiplied by the entire right-hand side of the original equation.

Note that the right-hand side contains the concentration of E at time T, $E(T)$, and the left-hand side contains only the concentration of E at time $T + 1$, $E(T + 1)$. This means that if we know the concentration of E at time T (i.e., the present), we can calculate the concentration of E at time $T + 1$ (i.e., the future), which is slightly ahead in time (Fig. 3.1). Therefore, if the initial value of E at time $T = 1$, $E(1)$, is given, then by repeating this calculation, we can sequentially calculate how the value of E changes thereafter (in MATLAB, we cannot set T=0, so we start from T=1). In other words, even if you don't know the exact solution, you can calculate and simulate the change in the value of E by increasing the value of T by 1 and repeating the calculation. This kind of calculation method is called the **Euler method**. For more details, please refer to more specialized books and materials, for example, [2, 3].

Let's actually do the calculation. In "script3_1A" below, dt=0.01, p=1, k=0.1, T varies from 1 to Tmax=5000, and the initial value of E at T=1 is 0. E is a vector with Tmax elements, and you can use *zeros* to prepare "E=zeros(1,Tmax);" or "E=zeros(Tmax,1);." It is better to prepare such a vector in advance. The former represents a single-row, Tmax-column matrix, and the latter a Tmax-row, single-column matrix. Either is fine, but since the vector has been a single-row matrix so far, we will follow the former notation (incidentally, "E=zeros (Tmax);" would be a square matrix with Tmax rows and Tmax columns).

Then the previous equation $E(T + 1) = (p - kE(T))dt + E(T)$ will look like "E (1,T+1) = (p - k * E (1,T)) * dt + E (1,T);" in MATLAB (Fig. 3.1). We can iterate this equation using the *for* statement while varying T from 1 to Tmax. Finally, you can use "plot(1:Tmax,E(1,1:Tmax));" to plot the time variation of E. If you put this idea into a program, the code will look like "script3_1A.m."

---

script3_1A.m

1	`Tmax=5000; dt=0.01;`	
2	`p=1; k=0.1;`	
3	`E=zeros(1,Tmax);`	
4	`for T=1:Tmax-1`	Increment T from 1 to Tmax−1
5	`  E(1,T+1)=(p-k*E(1,T))*dt+E(1,T);`	Calculate E according to the Euler method

(continued)

Fig. 3.2 Temporal change of the concentration of E (Euler method)

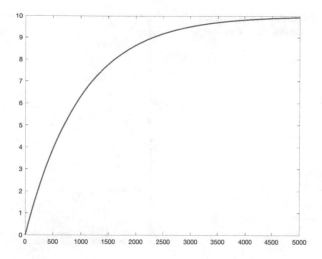

		Or, "E(T+1)=(p–k*E(T))*dt+E(T);"
6	end	
7	plot(1:Tmax, E(1,:));	Or, "plot(1:Tmax, E);"

The *for* statement in line 4 is set to "T=1:Tmax-1," because the left-hand side of line 5 becomes E(1,Tmax) exactly when T=Tmax–1. Also, since E is a vector matrix with only one row, we can omit the row information in this case. In other words, the fifth row can be written as "E(T+1)=(p-k*E(T))*dt+E(T);." In line 7, E(1,1:Tmax) is written as E(1,:), but it can also be written as "plot(1:Tmax, E);." The result of the calculation should look like Fig. 3.2.

Here, the value of E on the vertical axis increases as T on the horizontal axis increases and gradually (and asymptotically) approaches 10. The convergence of the values means that the value of E no longer changes with time. This means that $\frac{dE}{dt} = 0$, so if we apply this to Eq. (3.1.1), we get $p - kE = 0$, and thus $E = p/k = 10$, confirming that it should indeed asymptote to 10.

3.1.2 Comparison with Exact Solutions

Note that mathematically, the exact solution of *E* is

$$E(T) = \left(1 - e^{-k(T-1)}\right)\frac{p}{k} \qquad (3.1.2)$$

Using the exact solution, we can immediately find the value of $E(T)$ for any variable T, without having to do complicated calculations such as the Euler method. Substituting Eq. (3.1.2) into Eq. (3.1.1), the left-hand side becomes $\frac{dE}{dt} = p\, e^{-k(T-1)}$,

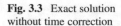

Fig. 3.3 Exact solution
without time correction

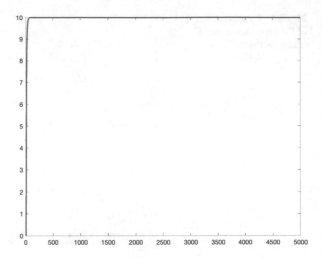

and the right-hand side becomes $p - kE = p - p(1 - e^{-k(T-1)}) = p\,e^{-k(T-1)}$,
indicating that this differential equation is indeed satisfied. When we have the exact
solution, we only need to substitute the value of T, so we don't need to do an
approximate calculation like the Euler method. Let's compare the result using the
exact solution with the Euler method.

script3_1B.m

1	`Tmax=5000;dt=0.01;`	
2	`p=1;k=0.1;`	
3	`T=1:Tmax;`	
4	`E=zeros(1,Tmax);`	
5	`E(1,:)=(1-exp(-k*(T-1)))*p/k;`	Calculate exact solution of E
6	`plot(1:Tmax,E);`	

Here, E is a vector, so "`E(1,:)`" in line 5 can simply be "E." Also, T is a vector
"`1:Tmax`", and "`(1-exp(-k*(T-1)))*p/k`" calculates the value of E at all
times using vector T. You can do the same calculations with the following
instructions:

```
for T=1:Tmax
   E(T)= (1-exp(-k*(T-1)))*p/k;
end
```

Here, the *for* statement is used to increase T by 1, but it is smarter to assign the
vector T directly to the exact solution without using the *for* statement.

The result is shown in Fig. 3.3, which is very different from Fig. 3.2. Note that the
time scale is actually quite different from that of "script3_1A.m" here. In the

Fig. 3.4 Exact solution
with time correction

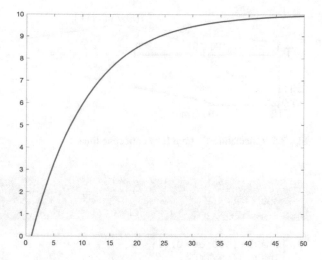

equation of Fig. 3.1, *dt* represents a small advance in time, so that even if time T advances by 1, time actually advances by only *dt* (dt=0.01 in "script3_1A.m"). Tmax represents the end time, but here time advances by only Tmax×dt=50. To make up for this difference, we need to multiply Tmax in "script3_1B.m" by *dt*. Now, let's use Tmax2 instead of Tmax by setting "Tmax2=Tmax*dt;." Then, the code will look like "script3_1C.m" below:

script3_1C.m

1	`Tmax=5000;dt=0.01;Tmax2=Tmax*dt;`	Calculate the end time, Tmax2
2	`p=1;k=0.1;`	
3	`T=1:Tmax2;`	
4	`E=zeros(1,Tmax2);`	
5	`E=(1-exp(-k*(T-1)))*p/k;`	
6	`plot(1:Tmax2,E);`	

This gives us the result in Fig. 3.4, which is very similar to Fig. 3.2. In Fig. 3.2, the time axis goes up to 5000 (=Tmax), which actually corresponds to 50 (=Tmax2).

It is a bit difficult to superimpose these two results and compare them directly. But if you are interested, you can try the following tasks. In "script3_1A," the time T is up to Tmax=5000, and in "script3_1C," it is up to Tmax2=50, so it is not possible to plot them simultaneously. As shown in Fig. 3.5, let's prepare the vector E1, which is the result of the calculation of E in "script3_1A," thinned out in the time direction, and plot it and the result of the exact solution E2 at the same time for comparison. "script3_1D.m" is the code:

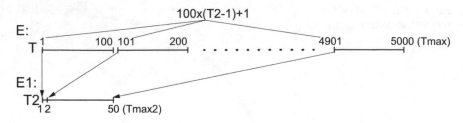

Fig. 3.5 Calculating E1 from E by correcting time

	script3_1D.m	

1	`Tmax1=5000;dt=0.01;` `Tmax2=Tmax1*dt;`	
2	`p=1;k=0.1;`	
3	`E=zeros(1,Tmax1);`	Prepare a vector E to store the solution by Euler method
4	`for T=1:Tmax1-1`	Calculate E by Euler method with end time Tmax1
5	`E(T+1)=(p-k*E(T))*dt+E` `(T);`	
6	`end`	
7	`E1=zeros(1,Tmax2);`	Prepare a vector E1 to store the values of E thinned out in the time direction
8	`for T2=1:Tmax2`	Calculate E1 with E thinned out in time direction with end time Tmax2
9	`E1(T2)=E((T2-1)/dt+1);`	Map E on E1 according to Fig. 3.5
10	`end`	
11	`T2=1:Tmax2;`	Prepare a time vector T2 with the end time Tmax2
12	`E2=zeros(1,Tmax2);`	Prepare a vector E2 to store the exact solution
13	`E2(1,:)=(1-exp(-k*` `(T2-1)))*p/k;`	Calculate the exact solution E2 using T2
14	`plot(T2,E1,'ro',T2,` `E2,'b*');`	Plot the numerical solution E1 and the exact solution E2

In lines 7–10, we compute E1, which is the result of E thinned out in the time direction. As with E2, the time is up to Tmax2, so the *for* statement in line 8 is "for T2=1:Tmax2." In line 9, we set the time to "(T2-1)/dt+1," which gives us 1 when T2=1, 101 when T2=2, and 4901 when T2=Tmax2=50, as shown in Fig. 3.5. In line 14, the Eulerian value of E1 is drawn as a circle (red) and the exact solution value of E2 is drawn as an asterisk (blue) so that we can distinguish between the two points even if they completely overlap. The result should look like Fig. 3.6. As you can see, the Euler method gives almost the same result as the exact solution.

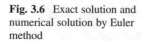

Fig. 3.6 Exact solution and numerical solution by Euler method

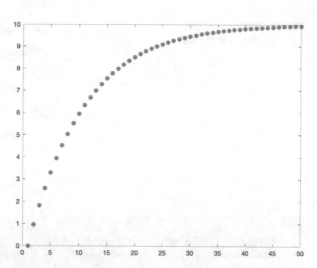

However, please keep in mind that the numerical solution is only an approximate solution and not the exact solution. In some cases, the error between the two solutions will be much larger. You may think that there is no need for numerical solutions, but the problem is that exact solutions cannot be obtained for most differential equations. So, in reality, it is important to use a computer to calculate the numerical solution.

The right-hand side of Eq. (3.1.1) is "$p - kE$," which is a linear function of E. A differential equation consisting of such a **linear function** is called a **linear differential equation**. Also, here we are not considering **differentiation** in the spatial direction, but only in the time direction.[1] A differential equation that contains only the derivative of one variable is called an **ordinary differential equation**. The differential equation treated here is called a **linear ordinary differential equation**, which is the easiest differential equation to handle, and the exact solution can be obtained by hand calculation. In the next section, we will discuss another example of a linear ordinary differential equation.

In any case, it is important to first write the equation properly, as in Eq. (3.1.1), and then transform the equation into a form that the computer can understand, such as "E(T+1) = (p-k*E(T))*dt+E(T) ;." If you omit this process before you are familiar with it, you may make a terrible mistake and end up with a completely meaningless calculation. For example, if you misplace the parentheses and write "E (T+1) = (p-k*E(T)*dt)+E(T) ;" you will be in big trouble. If you remember that "*dt* **hangs over the entire right-hand side of the original expression**," you should be fine in the scope of this book.

---

[1] Note that in mathematics, "differentiation" means the process of finding a derivative of a function. On the other hand, in biology, "differentiation" means the process in which an undifferentiated cell such as a stem cell becomes a differentiated cell type with specific characteristics. When equations are the topic, it means the former. When cells are the topic, it means the latter.

Fig. 3.7 Self-renewal and
differentiation of stem cells

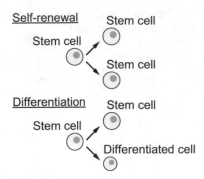

Note 3. Common Bugs #2: Grammatical Errors

The commands and functions used in MATLAB have a defined usage, or grammar, and if your program does not conform to that grammar, you will get an error message. For example, if you change "`plot(X,Y);`" to "`plot(X, Y,T);`" in "script2_2B.m" line 4, you will get the following message:

```
Error using plot
Data must be a single matrix Y or a list of pairs X,Y.
Error in script2_2B (line 4)
figure; plot(X,Y,T);
```

It says that there is a problem in line 4 because you used three vectors for *plot*. "`plot3(X,Y,T);`" would have no problem because *plot3* uses three vectors to draw a 3D graph instead of *plot*.

As you can see, grammatical errors usually cause error messages, so you should check the message and the syntax of the command that caused the error in the MATLAB documentation to find the cause of the error.

3.2 Mathematical Model of Hematopoietic Stem Cells

3.2.1 Mathematical Model of Hematopoietic Stem Cells: Part 1

In the previous section, we dealt with the synthesis and degradation of proteins, but differential equations can also be used to describe the cellular level phenomena of **cell proliferation** and **differentiation**[2]. Our bodies are composed of many different types of cells, and in many cases, special cells called stem cells give rise to these various types of cells. Stem cells **self-renew** to produce and proliferate cells that are identical to themselves, but they also produce differentiated cells that have different characteristics from stem cells (Fig. 3.7) [4]. These differentiated cells also

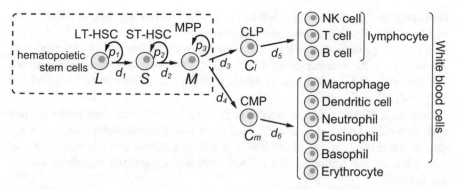

Fig. 3.8 Production of the blood cell system from the hematopoietic stem cells

repeatedly proliferate and differentiate, resulting in the production of various types of differentiated cells, which in turn build our brains, livers, and blood cell systems.

Let's consider the proliferation and differentiation of hematopoietic stem cells. All the cells of the blood cell system, such as red blood cells and white blood cells, are produced from hematopoietic stem cells (HSCs) in the bone marrow (Fig. 3.8) [5]. Among HSCs, Long-Term HSCs (LT-HSCs), which exist in the bone marrow for a long time, act as the top stem cells. LT-HSCs produce Short-Term HSCs (ST-HSCs), which are slightly differentiated from LT-HSCs, and ST-HSCs produce Multipotent Progenitors (MPPs), which are also slightly differentiated from ST-HSCs. MPPs produce Common Lymphoid Progenitor (CLP), a progenitor of leukocytes, and Common Myeloid Progenitor (CMP), a progenitor of myeloid cells [5]. Various blood cell lineages are generated from CLP and CMP to construct our blood cell lineage. Therefore, in the treatment of leukemia, it is expected that all the blood cells will be restored by transplanting LT-HSCs into the bone marrow after the artificial death of the blood cells.

The types of cells that can ultimately be produced by HSCs are extremely diverse, and the mechanisms that give rise to this diversity have yet to be fully elucidated. Here, we first consider a mathematical model of the differentiation of LT-HSCs into ST-HSCs and MPPs (dashed lines in Fig. 3.8).

Since LT-HSCs reside in the bone marrow and more differentiated cells are transported throughout the body by the bloodstream, the spatial distribution of these cells should be taken into account. However, this would result in a very complicated mathematical model. Here, we will ignore such spatial distribution and consider a mathematical model of ordinary differential equations that takes into account the total amount of cells in the whole body.

Here, L is the number of LT-HSCs, S is the number of ST-HSCs, and M is the number of MPPs. The unit time is 1 day, and LT-HSCs proliferate at a rate of p_1 per day and differentiate into ST-HSCs at a rate of d_1 per day. Since LT-HSCs cannot proliferate or differentiate in the absence of LT-HSCs, both of these rates depend on the number of LT-HSCs (L) at that time. Therefore, the time difference of L is calculated as $\frac{dL}{dt} = p_1 L - d_1 L$. When LT-HSCs differentiate into ST-HSCs, the number of LT-HSCs decreases at a rate of $-d_1 L$.

ST-HSCs proliferate at a rate of p_2 and differentiate into MPPs at a rate of d_2 depending on the number of ST-HSCs (S). Considering the increase due to the differentiation of LT-HSCs into ST-HSCs, $d_1 L$, the time difference of S is calculated as $\frac{dS}{dt} = d_1 L + p_2 S - d_2 S$.

Similarly, MPPs proliferate at a rate of p_3 and differentiate into other cells at a rate of d_3 depending on the number of MPPs (M), and we also consider the increment $d_2 S$ due to the differentiation of ST-HSCs into MPPs. Then, a mathematical model consisting of the following three linear ordinary differential equations can be constructed:

$$\frac{dL}{dt} = p_1 L - d_1 L \tag{3.2.1a}$$

$$\frac{dS}{dt} = d_1 L + p_2 S - d_2 S \tag{3.2.1b}$$

$$\frac{dM}{dt} = d_2 S + p_3 M - d_3 M \tag{3.2.1c}$$

These equations can be transformed according to the Euler method described in Sect. 3.1.1 and converted to MATLAB format as follows. Please write the equations on paper and check them yourself. In particular, please pay attention to the position of parentheses. Note that subscripts and italics are not allowed in MATLAB, so, for example, p_1 should be written with the usual combination of letters and numbers, as p1.

```
L(T+1)=(p1*L(T)-d1*L(T))*dt+L(T);
S(T+1)=(d1*L(T)+p2*S(T)-d2*S(T))*dt+S(T);
M(T+1)=( d2*S(T)+p3*M(T)-d3*M(T))*dt+M(T);
```

There are some studies in which each parameter is estimated from measured values, so let's use those values as a reference and set p1=0.009, d1=0.009, p2=0.042, d2=0.045, p3=4, d3=4.014 [5]. Let us assume that the initial values at time T=1 are L(1)=1, S(1)=0, and M(1)=0. In the *plot* in line 13 ("script3_2A.m"), L is plotted as a black line, S as a green line, and M as a magenta line. Did you get results like those in Fig. 3.9? The number of LT-HSC (L) does not change, but the number of ST-HSC (S) increases with time. Although L, S, and M are the number of cells, the results are not integers but include decimal values. This does not have to be interpreted as the exact number of cells, but rather as the average number of cells measured several times, or the density in the body.

Fig. 3.9 Proliferation of LT-HSC, ST-HSC, and MPP

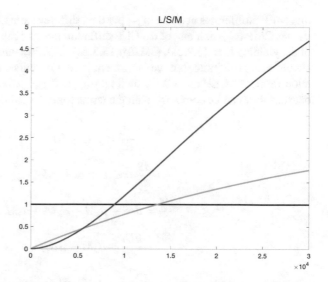

	script3_2A.m	
1	`dt=0.01;Tmax=30000;`	
2	`L=zeros(1,Tmax);L(1)=1;`	LT-HSC (initial value 1)
3	`S=zeros(1,Tmax);S(1)=0;`	ST-HSC (initial value 0)
4	`M=zeros(1,Tmax);M(1)=0;`	MPP (initial value 0)
5	`p1=0.009;p2=0.042;p3=4;`	
6	`d1=0.009;d2=0.045;d3=4.014;`	
7	`for T=1:Tmax-1`	
8	`  L(T+1)=L(T)+dt*(p1-d1)*L(T);`	
9	`  S(T+1)=S(T)+dt*(d1*L(T)` `+(p2-d2)*S(T));`	
10	`  M(T+1)=M(T)+dt*(d2*S(T)` `+(p3-d3)*M(T));`	
11	`end`	
12	`X=1:Tmax;`	
13	`plot(X,L,'k',X,S,'g',X,M,'m');`	Plot L, S, M with black, green, and magenta lines

If you want to add a legend in the graph, there is a *legend* command. Try adding "`legend('L','S','M');`" after line 13. You can add descriptions of the graphs L, S, and M to the previous *plot*.

3.2.2   Mathematical Model of Hematopoietic Stem Cells: Part 2

MPP gives rise to CLP, the progenitor of leukemic cells, and CMP, the progenitor of myeloid cells (Fig. 3.8). Modifying Eqs. (3.2.1a), (3.2.1b), and (3.2.1c), we assume that MPP proliferates at a rate of p_3 per day, differentiating into CLP (C_l) at a rate of d_3 and CMP (C_m) at a rate of d_4. CLP differentiates at a rate of d_5 and CMP at a rate of d_6, yielding Eqs. (3.2.2a), (3.2.2b), (3.2.2c), (3.2.2d), and (3.2.2e). CLP and CMP should also proliferate, but we omit them here. Of course, it is easy to add proliferation terms for CLP and CMP, such as p_4C_l and p_5C_m, but they are not important because they can be combined with the terms for d_5C_l and d_6C_m. We omit them here.

$$\frac{dL}{dt} = p_1L - d_1L \tag{3.2.2a}$$

$$\frac{dS}{dt} = d_1L + p_2S - d_2S \tag{3.2.2b}$$

$$\frac{dM}{dt} = d_2S + p_3M - d_3M - d_4M \tag{3.2.2c}$$

$$\frac{dC_l}{dt} = d_3M - d_5C_l \tag{3.2.2d}$$

$$\frac{dC_m}{dt} = d_4M - d_6C_m \tag{3.2.2e}$$

In "script3_2B.m" below, we will calculate Eqs. (3.2.2a), (3.2.2b), (3.2.2c), (3.2.2d), and (3.2.2e). Please check the transformation of the equation based on the Euler method by yourself. Although the values are not based on actual measurements, let's set d_3=0.022, d_4=3.992, d_5=0, d_6=0.5 as parameters to match the observations, and the initial values are L(1)=1, S(1)=0, M(1)=0, C_l(1)=0, C_m(1) =0.

If we write L, S, M, C_l, and C_m in one graph, it will be very busy, so we would like to display them separately in two graphs, 'L, S, M' and 'L, C_l, C_m'. The *subplot* command in lines 17 and 18 is used for the first time. It is used to tile multiple graphs in one window. Use *subplot* prior to *plot* as follows:

"subplot(number of tiles vertically, number of tiles horizontally, number of tiles to draw);"

The number of vertical and horizontal tiles is specified by "number of tiles vertically" and "number of tiles horizontally," respectively. The position of the actual tile to draw is specified by "number of tiles to draw." In line 17, "subplot(1,2,1);" specifies that the graph is to be drawn on the first (left) tile of the tiles with one vertical row and two horizontal columns, and "subplot(1,2,2);" in line 18 instructs us to draw on the second (right) tile.

The *figure* command in line 16 "figure('Position',[0 300 1000 400]);" has been used before, but it prepares a new window for us. Here, we will have two panels side by side in one window, so we need a horizontal window to make it look good. 'Position' declares that you will specify the location and size of the window, followed by the numerical value of "[leftmost coordinate bottom coordinate width height]." "[0 300 1000 400]" means that the window will be located 0 pixels from the left and 300 pixels from the bottom of the monitor, with a width of 1000 pixels and a height of 400 pixels. Since these values vary depending on the resolution of your PC monitor, it is better to draw a new window in the command window using "figure('Position',[leftmost coordinate, bottom coordinate, width, height]);" and try changing the values by yourself.

---

script3_2B.m

1	`dt=0.01;Tmax=30000;`	
2	`L=zeros(1,Tmax);L(1)=1;`	LT-HSC (initial value 1)
3	`S=zeros(1,Tmax);S(1)=0;`	ST-HSC (initial value 0)
4	`M=zeros(1,Tmax);M(1)=0;`	MPP (initial value 0)
5	`Cl=zeros(1,Tmax);Cl(1)=0;`	Lymphocyte common progenitor cell (initial value 0)
6	`Cm=zeros(1,Tmax);Cm(1)=0;`	Myeloid common progenitor cell (initial value 0)
7	`p1=0.009; p2=0.042; p3=4;`	
8	`d1=0.009; d2=0.045; d3=0.022;` `d4=3.992; d5=0.00; d6=0.5;`	
9	`for T=1:Tmax-1`	
10	`  L(T+1)=L(T)+dt*(p1-d1)*L(T);`	
11	`  S(T+1)=S(T)+dt*(d1*L(T)+(p2-d2)*S` `(T));`	
12	`  M(T+1)=M(T)+dt*(d2*S(T)+(p3-d3-d4)` `*M(T));`	
13	`  Cl(T+1)=Cl(T)+dt*(d3*M(T)-d5*Cl` `(T));`	
14	`  Cm(T+1)=Cm(T)+dt*(d4*M(T)-d6*Cm` `(T));`	
15	`end`	
16	`X=1:Tmax; figure('Position',[0 300` `1000 400]);`	
17	`subplot(1,2,1);plot(X, L, 'k', X, S,` `'g', X, M, 'm');`	
18	`subplot(1,2,2);plot(X, L, 'k', X, Cl,` `'b', X, Cm, 'r');`	

In the left panel of Fig. 3.10, L is plotted in black, S in green, and M in magenta. In the right panel, L is plotted in black, Cl in blue, and Cm in red. If you want to add a legend using *legend*, you can use "legend('L','S','M');" after line 17 and "legend('L','Cl','Cm');" after line 18.

Equations (3.1.1), (3.2.1a), (3.2.1b), (3.2.1c), (3.2.2a), (3.2.2b), (3.2.2c), (3.2.2d), and (3.2.2e) contain only time derivatives and the right-hand sides are all linear. The exact solution of a linear ordinary differential equation can be obtained by hand calculation. However, the **nonlinear ordinary differential equations** that we will deal with in the next and subsequent sections are not simple linear functions, and

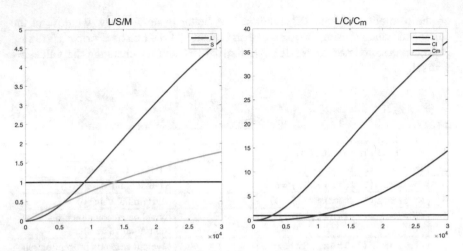

Fig. 3.10 Proliferation of LT-HSC, ST-HSC, MPP, CLP, and CMP

exact solutions are rarely obtained, so numerical calculations by computer are more important.

3.2.3   Logistic Equation

Since the nonlinear terms used in this section are a bit complicated, you can skip this section if you want. Let's review Eq. (3.2.1a), $\frac{dL}{dt} = p_1 L - d_1 L$. Since we set the parameters $p_1 = 0.009$ and $d_1 = 0.009$, the right-hand side of this equation is zero, and the value of L is constant, as can be seen in Fig. 3.9. If $p_1 < d_1$, the right-hand side of the equation is negative as long as L is positive, so the value of L will continue to decrease and become closer to zero. On the other hand, if $p_1 > d_1$, the value of L will keep increasing (when $L > 0$). For example, if we set the parameters of "script3_2A. m" as p1=0.02, p2=0.06, p3=4.01, d1=0.01, d2=0.05, d3=4, and the proliferation rate is always larger than the differentiation rate, all cells will increase as much as possible.

In reality, however, this is not possible because of the limited size of the human body and the limited amount of nutrients that can be ingested (such a condition can be called cancer). In reality, there is an upper limit to the density of cells due to spatial and nutritional limitations. The equation that is often used to set such an upper limit is called a **logistic equation**.

In the above equation, $p_1 L$ is the term that represents the proliferation of L. Here, this term is modified to $p_1 L\left(1 - \frac{L}{L_{max}}\right)$, so that when L approaches the upper limit, L_{max}, no further proliferation is allowed. In general, $\frac{dL}{dt} = pL\left(1 - \frac{L}{L_{max}}\right)$ is called the logistic equation for L (where p_1 is replaced with p), and when L is small enough,

L increases because the parentheses are close to 1, and the closer L is to L_{max}, the closer the parentheses are to 0 and L hardly increases.[2] This type of multiplication is called **logistic growth**. When this is applied to Eqs. (3.2.1a), (3.2.1b), and (3.2.1c), the result is as follows:

$$\frac{dL}{dt} = p_1 L \left(1 - \frac{L}{L_{max}} \right) - d_1 L \tag{3.2.3a}$$

$$\frac{dS}{dt} = d_1 L + p_2 S \left(1 - \frac{S}{S_{max}} \right) - d_2 S \tag{3.2.3b}$$

$$\frac{dM}{dt} = d_2 S + p_3 M \left(1 - \frac{M}{M_{max}} \right) - d_3 M \tag{3.2.3c}$$

If you write the code with $L_{max} = S_{max} = M_{max} = 5$, it will be as follows. Please run it yourself to check the result. It is true that the values of L, S, and M do not become infinitely large, but they are not necessarily close to 5 because of the influence of terms other than logistic multiplication.

script3_2C.m

1	`dt=0.01;Tmax=30000;`	
2	`L=zeros(1,Tmax);L(1)=1;`	
3	`S=zeros(1,Tmax);S(1)=0;`	
4	`M=zeros(1,Tmax);M(1)=0;`	
5	`p1=0.02;p2=0.06;p3=4.01;`	
6	`d1=0.01;d2=0.05;d3=4;`	
7	`Lmax=5;Smax=5;Mmax=5;`	
8	`for T=1:Tmax-1`	
9	`  L(T+1)=L(T)+dt*( p1*L(T)*(1-L(T)/Lmax)-d1*L(T) );`	
10	`  S(T+1)=S(T)+dt*( d1*L(T)+p2*S(T)*(1-S(T)/Smax)-d2*S(T) );`	
11	`  M(T+1)=M(T)+dt*( d2*S(T)+p3*M(T)*(1-M(T)/Mmax)-d3*M(T) );`	
12	`end`	
13	`X=1:Tmax;`	
14	`plot(X,L,'k',X,S,'g',X,M,'m');`	Plot L in black, S in green, and M in magenta

---

[2] The logistic equation has an exact solution, and if the initial value of L at $t=0$ is L_0, the solution to $\frac{dL}{dt} = pL \left(1 - \frac{L}{L_{max}} \right)$ is $L = \frac{L_{max}}{1+(L_{max}/L_0)e^{-pt}}$.

Fig. 3.11 Relationship
between glucose, insulin,
and diet

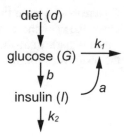

Exercise 3.2.3a Calculate the convergence of L, S, and M by applying $\frac{dL}{dt} = \frac{dS}{dt} = \frac{dM}{dt} = 0$ to Eqs. (3.2.3a), (3.2.3b), and (3.2.3c).

Exercise 3.2.3b Compare the numerical (Eulerian) and exact solutions of Eq. (3.2.3a) by superimposing them in a *plot* graph ($L_0 = 1$, $L_{max} = 5$; see "script3_1D.m").

3.3   Mathematical Model of Glucose Metabolism

Our **blood glucose level**, or the concentration of **glucose** in the blood, is kept almost constant, and this phenomenon can be understood mainly from the relationship between blood glucose level and **insulin** [6]. Insulin is a hormone that lowers the blood glucose level, and even if it rises transiently due to eating, insulin is secreted in response to the blood glucose level to lower it again. The mechanism of controlling blood glucose level is very complicated, but let us consider a simple mathematical model that focuses only on the relationship between diet (d), glucose concentration (G), and insulin concentration (I).

First, let's consider the change in blood glucose concentration over time, $\frac{dG}{dt}$. Glucose concentration increases with diet, d (where d is the rate of change per unit time of glucose concentration in the body that increases with diet). Let's simply assume that glucose is supplied by the food intake and simply add d. Glucose is broken down through a very complex biochemical process, and insulin regulates multiple steps in this process [6]. In this book, we do not want to completely describe the breakdown of glucose and its regulation by insulin in mathematical terms but rather use simple equations.

If we assume that glucose is broken down at a rate of k_1 per unit of time (let's say 1 hour), then $\frac{dG}{dt} = d - k_1 G$. The important point is that this breakdown of glucose is accelerated by insulin (Fig. 3.11). How can insulin increase the rate of glucose breakdown? In the following mathematical model, the degradation of glucose is assumed to be $-k_1 G(1 + aI)$. In the absence of insulin and with $I = 0$, glucose is always broken down spontaneously at a rate of k_1, but with $(1 + aI)$, the presence of insulin further increases the rate of glucose breakdown. a is a coefficient that corrects the biochemical reaction of glucose according to the concentration of insulin. On the other hand, if we consider that insulin is secreted at a rate of b per unit time in proportion to the concentration of glucose and is broken down at a rate of k_2, then $\frac{dI}{dt} = bG - k_2I$. These can be summarized as:

$$\frac{dG}{dt} = d - k_1G(1 + aI) \qquad (3.3.1a)$$

$$\frac{dI}{dt} = bG - k_2I \qquad (3.3.1b)$$

It is important to note that in Eq. (3.3.1a), the degradation of glucose is $-k_1G(1 + aI)$, which is different from the previous equations. The presence of the G times I term makes this equation more than just a linear function of G. It is a nonlinear differential equation. Of course, if the equation is a quadratic or cubic function of G, it is also called a nonlinear differential equation. However, since the Eulerian transformation is the same for both linear and nonlinear equations, we do not need to be particularly conscious of it when programming.

Let's assume that a=b=1, k1=k2=0.1, dt=0.01, and Tmax=10000 (dt×Tmax=100 h). These values are not based on actual measurements but are arbitrary. The values obtained by the program do not correspond directly to actual blood glucose or insulin concentrations, and we assume that we are discussing the relative relationship between blood glucose and insulin concentrations.

Although d seems to be a constant, in reality, we are not always eating, but we are eating at certain times of the day. For simplicity, let's assume that we take a meal every 10 h for 1 h (interval between meals, $dint = 1000$). Now, let's prepare a vector d with T_{max} elements, "dint=1000; d=zeros(1,Tmax);." If we take a meal of unit amount 1 from time "dint-100 to dint" between "time 1 to dint," then "d(1, dint-100:dint)=1;." This is repeated ten times as Tmax/dint=10, so "d=repmat(d(1,1:dint),1,fix(Tmax/dint));" (see Sect. 2.10.2 for *repmat*). In case Tmax is not divisible by dint, we truncate the decimal point with "fix(Tmax/dint)." Then, the following "script3_3A.m" is obtained:

script3_3A

1	dt=0.01;Tmax=10000;dint=1000;	
2	d=zeros(1,Tmax);d(dint-100:dint)=1;	Or, "d(dint-100: dint)=0.5;"
3	d=repmat(d(1,1:dint),1,fix(Tmax/dint));	
4	G=zeros(1,Tmax);G(1)=1;	Glucose concentration
5	I=zeros(1,Tmax);I(1)=1;	Insulin concentration
6	a=1;b=1;k1=0.1;k2=0.1;	Or, "b=0" or "a=0"

(continued)

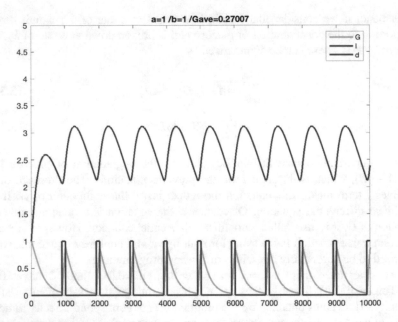

Fig. 3.12 Results of normal condition

7	for T=1:Tmax-1	
8	G(T+1)=G(T)+dt*(-k1*G(T)*(1+a*I(T)) + d(T));	
9	I(T+1)=I(T)+dt*(b*G(T) - k2*I(T));	
10	end	
11	Gave=sum(G(1,Tmax/2+1:Tmax))*2/Tmax;	
12	X=1:Tmax;	
13	plot(X,G,'g',X,I,'r',X,d,'b');ylim([0 5]);leg- end('G','I','d');	
14	title(strcat('a=',num2str(a),' /b=',num2str (b),' /Gave=',num2str(Gave)));	

The results are shown in Fig. 3.12. The green line shows the blood glucose level (*G*), which rises with diet (*d*), represented by the blue line. The red line shows the insulin concentration (*I*), which rises a little later. *G* decreases as *I* rises, and *I* decreases a little later. The cycle repeats multiple times.

Here, G_{ave}, which is calculated in line 11, is the average value of blood glucose, and since *sum* calculates the sum of the elements of a vector (see Sect. 1.5.1), we can obtain the average value of *G* over all periods by setting "Gave=sum(G)/Tmax." However, we only consider the second half of the whole period, when the rhythm becomes constant, because the values may transiently show erratic behavior right after the start of the calculation. In other words, to calculate the average of *G* from time 5001 to time 10000 (T=Tmax/2+1 to T=Tmax), we use "Gave=sum(G(1, Tmax/2+1:Tmax))*2/Tmax;." Here, Gave=0.27007 (Fig. 3.12). In line 14, we

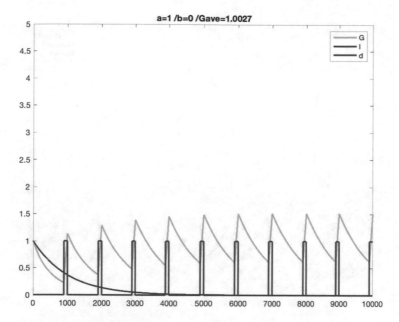

Fig. 3.13 Results of type 1 diabetes

use *title*, *strcat*, and *num2str* commands to write the values of a, b, and Gave into the title of the graph (see Sects. 1.6.2 and 2.6 for details).

Let's use this mathematical model to calculate what happens when you have **diabetes**. There are two major types of diabetes: **type 1 diabetes**, which is caused by the destruction of pancreatic β cells, which are insulin-secreting cells, and **type 2 diabetes**, which is caused by the failure in insulin action due to various causes. In Asian countries, type 2 diabetes is very common and is thought to be caused by an impairment of the blood glucose-lowering effect of insulin (**insulin resistance**) due to lifestyle factors such as overeating and lack of exercise, in addition to genetic factors.

In the case of type 1 diabetes, insulin is not secreted. If you want to consider the case of sudden onset of the disease, we can set b=0 (Fig. 3.11). Let's change "b=1" in line 6 to "b=0" in "script3_3A.m" and run the program. Then, the insulin concentration drops rapidly as shown in Fig. 3.13. As a result, the average blood glucose level rises to Gave=1.0027, which is about four times higher than normal.

To treat this abnormal blood glucose level, we can administer an insulin drug. But how can we modify the program to administer the insulin drug at the same time as the food intake? Let us assume that the insulin concentration obtained by the insulin drug is given by a vector *p* and its value is the same as *d*. In Eq. (3.3.1b), insulin is produced according to the blood glucose level, and we can add the term *p* here.

Let's add "p=d" to the third line of "script3_3A," "d=repmat(d(1,1:dint),1,fix(Tmax/dint));p=d;," and modify line 9 as "I(T+1)=I(T)

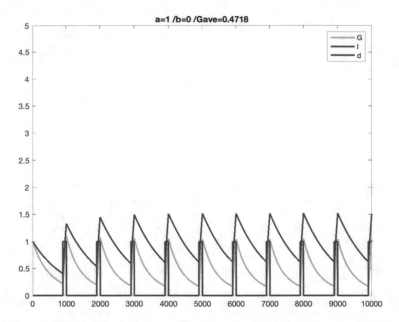

Fig. 3.14 Administrating insulin

$+dt*(b*G(T) - k2*I(T) +p(T));$." We can see that the blood glucose level improves and its average decreases to Gave=0.4718 as shown in Fig. 3.14.

In type 2 diabetes, the glucose concentration does not decrease even if insulin is present due to insulin resistance, so let's change "a=1" in line 6 of "script3_3A" to "a=0" and run it (Fig. 3.11). As shown in Fig. 3.15, Gave becomes 1.0093, which is four times higher than normal. The concentration of insulin is also abnormally elevated. This is due to the large amount of insulin produced by the elevated blood glucose level, G.

Dietary therapy is an effective treatment for type 2 diabetes. In order to limit the amount of food intake, let's modify the second line of "script3_3A.m" to "d=zeros(1,Tmax);d(1,dint-100:dint)=0.5;" and run it. We can see that the average blood glucose level improves and decreases to Gave=0.50532 as shown in Fig. 3.16. If the values of each variable and parameter are adjusted to match the actual measured values, we can use such a mathematical model to discuss the mechanism of type 1 and type 2 diabetes and strategies for treatment.

Note 4. Common Bugs #3: Inappropriate Handling of Numbers and Letters

In "script2_6A.m," we assigned random numbers between 0 and 10 to a 4-by-5 matrix M. To find out the element in the second row and third column of the matrix M, we use "M(2,3)." Each number in the parentheses (**array**

(continued)

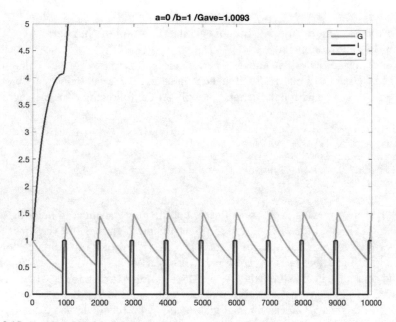

Fig. 3.15 Results of type 2 diabetes

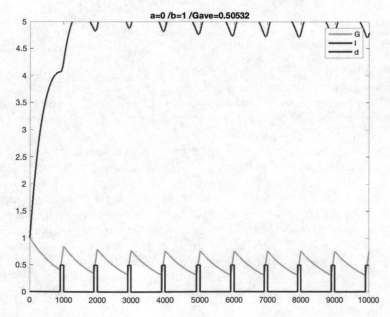

Fig. 3.16 Restricting diet

**Note 4. Common Bugs #3: Inappropriate Handling of Numbers
and Letters** (continued)
index) must be a positive integer. There cannot be "zero row" or "2.1 column"
when we refer to a matrix. So, if you change "for I=1:4" in line 2 to "for
I=1:0.1:4" and run the script, you will get the following error message:

```
Index in position 1 is invalid. Array indices must be positive
integers or logical values.
Error in script2_6A (line 4)
disp(strcat('Row', num2str(I), 'column', num2str(J), 'is ',
num2str(M(I,J)) ));
```

The error message appears in line 4, but the *disp* statement in line 4 is so
long that it is a little difficult to see where the problem is. However, since it
says "Array indices must be positive integers..." and the only
thing in line 4 that uses the array index is "num2str(M(I,J))," you can
find that this is the problem. Also, "The index at position 1 is
invalid" means that there is a problem with "I," the first index of "M(I,
J)." In this way, you can find out that there is a problem with the value of I.
    Also, if you try "for I=1:5" instead of "for I=1:4" in line 2, you will
get the following messages:

```
Index in position 1 exceeds array bounds (must not exceed 4).
Error in script2_6A (line 4)
disp(strcat('Row', num2str(I), 'column', num2str(J), 'is ',
num2str(M(I,J)) ));
```

The problem is that the index at position 1, that, "I," has exceeded the
number of rows in M, that is, 4. You can find that there is a problem with the
loop of the *for* statement in line 2.
    Similar to the previous grammatical error, there is also a problem when
letters and numbers are confused: if you don't use the *num2str* function in line
4 in "script2_6A.m" and change it to "disp(strcat('Row', I,
'column', J, 'is ', M(I,J)));," you will not get the error message.
But the message "Row column is" will be displayed repeatedly. The values
of I, J, and M(I,J) are supposed to be inserted in the text, but these values are
completely ignored. If you search for *strcat* in the MATLAB documentation,
you will see that this command is supposed to concatenate characters. So, if
you write a variable containing a number in parentheses after *strcat*, it will be
considered blank. We want these numbers to be considered as characters, so
we use *num2str* in "script2_6A.m."

(continued)

Note 4. Common Bugs #3: Inappropriate Handling of Numbers and Letters (continued)

In the above example, the problem was in the program itself, and we could find and fix it. But it is also possible that the problem is caused by a previously executed program. For example, run "script2_10B.m" first. Then, disable the first line of "script2_10C.m" by adding "%" as "%Z=zeros(100,100);" and run it. You will see that the result is completely different from Fig. 2.23. This is because "script2_10C.m" was executed with the matrix Z calculated in "script2_10B.m" still in the workspace. Lines 4–6 of "script2_10C.m" set "Z(Y,X)=1" only when X>=Y, so the original Z value remains in the workspace when X<Y.

To prevent this, you can reset the contents of the workspace by executing *clear* from the command window before executing "script2_10C.m," or you can reset the contents of Z by setting "Z=zeros(100,100);" in the first line of "script2_10C.m." This is why you prepare the matrices to be used in the program in advance using *zeros* and *ones* in other programs in this book. The matrices are stored in the memory of the computer, and preparing them at the beginning of the program has the advantage that the computer can process them more efficiently.

3.4 Mathematical Model of Infectious Disease

3.4.1 SIR Model

It is an important question how various epidemics spread in a population, how the number of infected people and those who recover from infection change over time, and what effect the use of antibiotics, vaccines, and isolation of the population has. **The SIR model**, a classical mathematical model of epidemics, is also useful in considering countermeasures for real problems [1, 3, 7].

In this section, we consider the relationship among three variables: *Susceptible* (*S*), *Infected* (*I*), and *Recovered* (*R*) (Fig. 3.17). Infection is a process in which both the uninfected and the infected are present, and if one of them is not present, infection does not happen and the number of infected people will not increase.

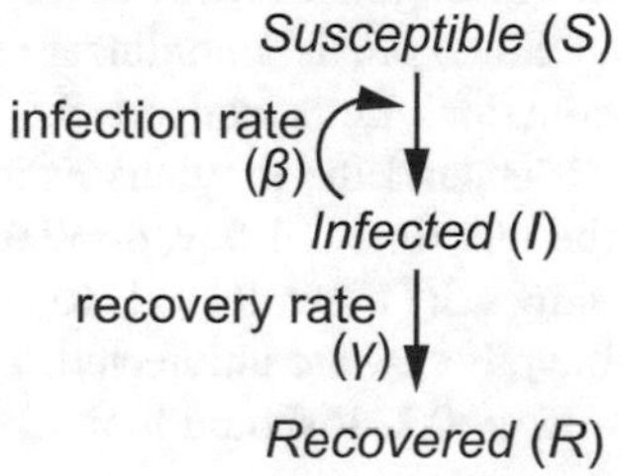

Fig. 3.17 Relationships between *S*, *I*, and *R*

Conversely, if both of them increase, the number of infected people will increase significantly. Since the number of new infected people increases when the uninfected and infected people meet, the rate of increase of the number of infected people I per unit time (let's say a day) is proportional to both the number of uninfected people S and the number of infected people I, that is, it is the product of S and I. The coefficient of the rate of infection is β (beta). In other words, the number of infected people increases at a rate of βSI per unit time.

On the other hand, infected people who have previously been infected will recover with time. This recovery process depends mainly on the immunity of the infected person and does not depend on the number of infected persons or the number of recovered persons. We define the recovery rate as γ (gamma) and assume that the number of infected people decreases at a rate of γI per unit time. Combining these two effects, we get $\frac{dI}{dt} = \beta SI - \gamma I$. Also, since the number of uninfected people decreases at the same rate as the rate at which uninfected people become infected, βSI, we have $\frac{dS}{dt} = -\beta SI$.

Finally, since the infected recover and the number of infected decreases at the rate of γI and the number of recovered increases at the rate of γI, we have $\frac{dR}{dt} = \gamma I$. Putting these three equations together, we get the following mathematical model, where the values of S, I, and R are not integers but include decimal places, which can be read as the average number of people measured several times rather than simply the number of people:

$$\frac{dS}{dt} = -\beta SI \tag{3.4.1a}$$

$$\frac{dI}{dt} = \beta SI - \gamma I \tag{3.4.1b}$$

$$\frac{dR}{dt} = \gamma I \tag{3.4.1c}$$

Note that adding the left and right sides of each of the three equations together yields $\frac{dS}{dt} + \frac{dI}{dt} + \frac{dR}{dt} = 0$. This means that the total population of the susceptible, infected, and recovered is unchanged. For example, even if the number of uninfected people decreases, the number of infected and recovered people increases, so the total population remains the same. The assumption that the total population will remain unchanged if there is no movement in and out of the population is valid. Note that R can mean the number of deaths as well as the number of recoveries, but here we do not distinguish between recoveries and deaths.

Since βSI is a nonlinear term, it can be called a nonlinear ordinary differential equation. Let's convert this to a program based on the Euler method. If you understand the previous examples, this should not be difficult. Let's assume that the unit time is 1 day, dt=0.01, Tmax=10000 (dt×Tmax=100 days), and the initial values S(1)=99, I(1)=1, R(1)=0 when T=1, that is, only one person is infected and the other 99 are uninfected. The parameters β (or, b) and γ (or, g) are set to b=0.01 and g=0.1. It should look like the following script (Fig. 3.18 left).

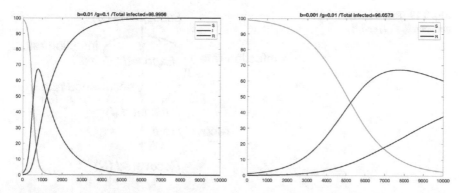

Fig. 3.18 Calculation examples of SIR model

script3_4A.m

1	`dt=0.01;Tmax=10000;`	
2	`S=zeros(1,Tmax);S(1)=99;`	Susceptible
3	`I=zeros(1,Tmax);I(1)=1;`	Infected
4	`R=zeros(1,Tmax);R(1)=0;`	Recovered
5	`b=0.01;g=0.1;`	Or, "b=0.001; g=0.01;"
6	`for T=1:Tmax-1`	
7	`    S(T+1)=S(T)+dt*( -b*S(T)*I(T) );`	
8	`    I(T+1)=I(T)+dt*( b*S(T)*I(T) -g*I(T) );`	
9	`    R(T+1)=R(T)+dt*( g*I(T) );`	
10	`end`	
11	`X=1:Tmax;`	
12	`plot(X,S,'g',X,I,'r',X,R,'b');legend` `('S','I','R');`	
13	`title(strcat('b=',num2str(b),' /g=',num2str` `(g),' / Total infected =',num2str(S(1)-S(Max)) ));`	

In line 12, S is plotted as a green line, I as a red line, and R as a blue line. If we set $\beta=0.001$ and $\gamma=0.01$, both the infection rate and the recovery rate will be reduced by a factor of 10. As a result, the overall time trend will be slower as shown in Fig. 3.18 right.

In line 13, the cumulative total number of infected people is calculated and displayed. Although the value of "$I(T)$" indicates the number of infected people at time T, it also decreases as the infected people recover. So, we cannot calculate the cumulative total number of infected people using I(T). However, when the number of susceptible people "$S(T)$" decreases with time, we can say that this decrease is the cumulative total number of infected people. So, we can calculate the cumulative

Fig. 3.19 Relationships
between *S*, *E*, *I*, and *R*

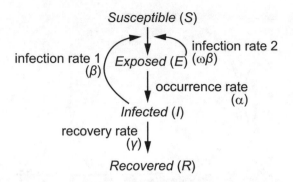

total number of infected people at time T as "S(1)-S(T)," and the cumulative total number of infected people at time Tmax, or the final cumulative total number of infected people, as "S(1)-S(Tmax)."

Exercise 3.4.1a In the case of COVID-19, the long incubation period after infection leads to the spread of the virus. Consider a four-variable model, that is, **SEIR model**, by adding the variable *Exposed* (*E*), which means that the patient is infected but not infectious (Fig. 3.19). Infection rate β (or, b)=0.001, occurrence rate α (or, a)=0.01, recovery rate γ (or, g)=0.01. S(1)=99, E(1)=0, I(1)=1, R(1)=0 when T=1. Perform a numerical simulation.

Exercise 3.4.1b In the case of COVID-19, *Exposed* people are thought to be partially infectious. Let's suppose that the infection rate of *Exposed* people is ωβ, where ω (or, w)=0.5 (Fig. 3.19). S(1)=99, E(1)=1, I(1)=0, R(1)=0 when T=1. Perform a numerical simulation.

3.4.2   Comparing the Results of Multiple Parameters

So, what would happen if antibiotics and special drugs improved the recovery rate γ? Of course, if the value of γ becomes very high, all infected people will quickly recover. On the other hand, such a mathematical model can be useful in figuring out how to administer a drug with limited effectiveness. Let's calculate it comprehensively by varying the value of γ little by little and compare the results side by side. If we give the value of γ as a vector "g=[0.1 0.2 0.5 1.0];," you can try different parameters by sequentially referring to the elements of the vector using the *for* statement. Let's try some calculations using a series of recovery rates as in "script3_4B.m."

script3_4B.m

1	`dt=0.01;Tmax=10000;`	
2	`S=zeros(1,Tmax);S(1)=99;`	Susceptible
3	`I=zeros(1,Tmax);I(1)=1;`	Infected
4	`R=zeros(1,Tmax);R(1)=0;`	Recovered
5	`X=1:Tmax;`	
6	`b=0.01;g=[0.1 0.2 0.5 1.0];`	Assign recovery rates to a four-element vector g
7	`figure('Position',[0 400 1400 300]);`	Adjust the size and position of the window accordingly
8	`for J=1:4`	Increment J from 1 to 4 and use different elements of g(J)
9	`  for T=1:Tmax-1`	
10	`    S(T+1)=S(T)+dt*( -b*S(T)*I(T));`	
11	`    I(T+1)=I(T)+dt*( b*S(T)*I(T) -g(J)*I(T) );`	
12	`    R(T+1)=R(T)+dt*( g(J)*I(T) );`	
13	`  end`	
14	`  subplot(1,4,J);plot(X, S, 'g', X, I, 'r', X, R, 'b');`	
	`    legend('S','I','R');`	Plot in the J-th tile of the tiles with one vertical row and four horizontal columns
15	`  title(strcat('b=',num2str(b),' /g=', num2str(g(J)),`	
	`    ' /Total infected=', num2str(S(1)-S(Tmax)) ));`	
16	`end`	Return to line 8

Note that in "script3_4A.m," g was a scalar, but here it is a vector. Therefore, in lines 11 and 12, g is now g(J).

The *subplot* in line 14 was explained in Sect. 3.2.2. "`subplot(1,4,J);`" tells the window to draw in the J-th tile of the one vertical row and four horizontal columns. Since the value of J varies from 1 to 4, four tiles are displayed, one at a time, starting from the left.

As shown in Fig. 3.20, when the recovery rate γ increases, the peak height of the number of infected people, I, decreases and it immediately decreases after the peak. The model does not distinguish between the recovery rate and the death rate, and an increase in R can be equivalent to an increase in the death rate. This also means that the number of infected people is unlikely to increase for highly lethal epidemics, that is, those with large values of γ.

Now, in the absence of antibiotics or special drugs, we can reduce the rate of infection by using masks and handwashing. The situation when the infection rate is changed can be calculated by preparing the infection rate β as a vector like the

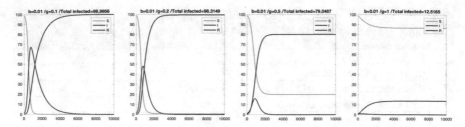

Fig. 3.20 Changing the recovery rate

following: "b=[0.01 0.005 0.002 0.001];." Then, the program will look like "script3_4.C.m."

script3_4C.m	

1	`dt=0.01;Tmax=10000;`	
2	`S=zeros(1,Tmax);S(1)=99;`	Susceptible
3	`I=zeros(1,Tmax);I(1)=1;`	Infected
4	`R=zeros(1,Tmax);R(1)=0;`	Recovered
5	`X=1:Tmax;`	
6	`b=[0.01 0.005 0.002 0.001];g=0.1;`	Assign infection rates to a four-element vector b
7	`figure('Position', [0 400 1400 300]);`	
8	`for J=1:4`	Increment J from 1 to 4 and use different elements of b(J)
9	`  for T=1:Tmax-1`	
10	`    S(T+1)=S(T)+dt*( -b(J)*S(T)*I` `(T) );`	
11	`    I(T+1)=I(T)+dt*( b(J)*S(T)*I` `(T) -g*I(T) );`	
12	`    R(T+1)=R(T)+dt*( g*I(T) );`	
13	`  end`	

14	`  subplot(1,4,J);plot(X,S,'g',X,`
	`I,'r',X,R,'b');legend`
	`('S','I','R');`	

15	`  title(strcat('b=',num2str(b`
	`(J)),' /γ=',num2str(g),`
	`    ' /Total infected=',num2str(S`
	`(1)-S(Tmax)) ));`	

16	`end`	Return to line 8

In "script3_4B.m," g is a vector, but here, g is a scalar and b is a vector. Therefore, in lines 10 and 11, b becomes "b(J)." Fig. 3.21 shows that as the rate of infection decreases, the number of infected people decreases, and the peak time when the number of infected people increases is delayed. In other words, even if there is no

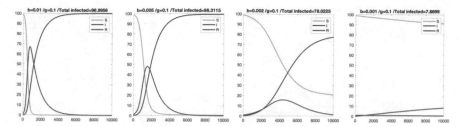

Fig. 3.21 Changing the infection rate

antibiotic or special medicine, the total number of infected people can be reduced if the rate of infection is lowered in some way over a long term.

In addition to improving the rate of infection β (or, b) and recovery γ (or, g), there is another important way to fight epidemics. It is to keep the population as small as possible. Let's prepare the total population N as a vector as follows: "N=[100 200 300 400];." To make the result easier to understand, let's set "b=0.001" and "g=0.01." Assuming that the initial number of infected people "I(1)" is always 1% of the total population N, the initial number of uninfected people "S(1)" is 99%, and the initial number of recovered people "R(1)" is 0%, the program will look like "script3_4D.m" below.

script3_4D.m

1	`dt=0.01;Tmax=10000;`	
2	`S=zeros(1,Tmax);`	Susceptible
3	`I=zeros(1,Tmax);`	Infected
4	`R=zeros(1,Tmax);R(1)=0;`	Recovered
5	`X=1:Tmax;`	
6	`b=0.001;g=0.1;`	
7	`N=[100 200 300 400];`	Prepare the values of N as a four-element vector
8	`figure('Position', [0 400 1400 300]);`	
9	`for J=1:4`	Increment J from 1 to 4
10	`  I(1)=N(J)*0.01;`	Initial number of infected people I(1)
11	`  S(1)=N(J)-I(1);`	Initial number of uninfected people S(1)
12	`  for T=1:Tmax-1`	
13	`    S(T+1)=S(T)+dt*( -b*S(T)*I(T) );`	
14	`    I(T+1)=I(T)+dt*( b*S(T)*I(T) -g*I(T) );`	
15	`    R(T+1)=R(T)+dt*( g*I(T) );`	
16	`  end`	

(continued)

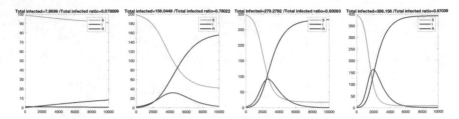

Fig. 3.22 Changing population

17	`subplot(1,4,J); plot(X,S,'g' X,I,'r', X,R,'b');`	
	`    ylim([0 N(J)]);legend ('S','I','R');`	
18	`title(strcat('Total infected=', num2str(S(1)-S(Tmax)),`	
	`    '/Total infected ratio=',num2str((S (1)-S(Tmax))/N(J)) ));`	
19	`end`	Return to line 9

In the *for* statement loop starting from line 9, the total population "N(J)" changes through 100, 200, 300, 400, etc. The initial number of infected people at T=1 can always be 1 (i.e., I(1)=1), but here the initial number of infected people is 1% of the population (i.e., I(1)=N(J)*0.01). The initial number of uninfected people can then be calculated as "S(1)=N(J)-I(1)." Since the initial values of S and I depend on the values of J and N, they are set in lines 10 and 11, that is, inside the *for* statement loop for J. This is important because until now, the initial values of S and I were set in lines 2 and 3, because these values were always unchanged. However, in "script3_4D.m," the initial values of S and I change whenever the value of N changes, so it is necessary to set them immediately after "for J=1:4" in line 9 instead of lines 2 and 3.

The result is shown in Fig. 3.22. We can see that the larger the population of the group, the more likely the number of infected people will increase, and the smaller the population of the group, the less likely the number of infected people will increase. From this, we can see that it is better to avoid large gatherings in order to prevent the spread of infectious diseases.

Exercise 3.4.2a Compare the results when the value of ω is provided as a vector "w=[0.1 0.5 0.9];" based on the SEIR model in Exercise 3.4.1b. a=0.01, b=0.001, g=0.01.

Exercise 3.4.2b Compare the results when the value of α is provided as a vector "a=[0.001 0.01 0.1];" based on the SEIR model in Exercise 3.4.1b. b=0.001, g=0.01, w=0.5.

3.4.3 Introducing the Basic Reproduction Number, R_0

When discussing the infectivity of an infectious disease, a value called the **basic reproduction number (R_0)** is very important. It refers to how many people on average an infected person is likely to spread the infection to. If $R_0 = 1$, there will be no sudden epidemic but no end to the disease. If $R_0 > 1$, it means that the infection may spread rapidly.

How should we deal with R_0 in the SIR model? In general, in the early stages of an epidemic, $R_0 = \frac{\beta}{\gamma}$. If the infection rate β is large, R_0 will be large, and if the recovery rate γ is large, R_0 will be small. In all of the above examples, $R_0 < 1$, so the simulations were for less infectious cases. Here, let's assume that $R_0 = 2$. Note that the values of β and γ can take many different values even when the value of R_0 is determined. Let's try four different values of β (or, b); "b=[0.0001 0.001 0.01 0.1];" as in "script3_4C.m." Then, since $\gamma = \frac{\beta}{R_0}$, we can find the value of g according to the value of b. Let's assume that the total population is N=500, the initial number of infected people is 1% of that, that is, "I(1)=N/100," and the initial number of uninfected people is "S(1)=N-I(1)." Then the program will look like "script3_4E.m" below:

	script3_4E.m	
1	dt=0.01;Tmax=10000;N=500;	
2	I=zeros(1,Tmax);I(1)=N/100;	Initial value of I
3	S=zeros(1,Tmax);S(1)=N-I(1);	Initial value of S
4	R=zeros(1,Tmax);R(1)=0;	Initial value of R
5	X=1:Tmax; R0=2;	
6	b=[0.0001 0.001 0.01 0.1];	Prepare the values of b as a four-element vector
7	figure('Position', [0 400 1400 300]);	
8	for J=1:4	Increment J from 1 to 4
9	g=b(J)/R0;	Calculate the value of g from b (J) and R0
10	for T=1:Tmax-1	
11	S(T+1)=S(T)+dt*(-b(J)*S(T)*I(T));	
12	I(T+1)=I(T)+dt*(b(J)*S(T)*I(T) -g*I(T));	
13	R(T+1)=R(T)+dt*(g*I(T));	
14	end	
15	subplot(1,4,J); plot(X,S,'g',X,I,'r',X,R,'b');	

(continued)

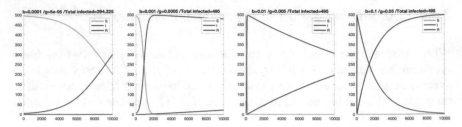

Fig. 3.23 Fixing the basic reproduction rate R_0 to 2

16	`title(strcat('b=',num2str(b(J)),' /` `g=',num2str(g),` `'/Total infected=',num2str(S` `(1)-S(Tmax)) ));`	
17	`end`	Return to line 8

Since the initial number of infected people, "`I(1)`," is always 1% of the population, we set "`I(1)=N/100`" in line 2. The initial number of uninfected people, "`S(1)`," is calculated based on "`I(1)`," so in line 3, "`S(1)=N-I(1)`."

Since b is a vector and g is a scalar, we need to calculate the value of g from b. We need to calculate the value of g every time the value of J changes in the *for* statement loop from line 8. So, we set "`g=b(J)/R0`" in line 9. Until now, the value of g did not always change. However, in "script3_4E.m," the value of g changes when the value of b(J) changes, so it is necessary to set g immediately after "`for J=1:4`" in line 8.

The result is shown in Fig. 3.23, and we can see that even if we say $R_0 = 2$, there is a considerable range of time variation in the number of infected people depending on the values of β and γ.

3.5   Lotka-Volterra Model

3.5.1   Numerical Calculation Using the Euler Method

Finally, let's deal with **Lotka-Volterra model** used in the field of ecology [1, 8, 9]. In this model, the relationship between the population densities of **prey (herbivore)** and **predators (carnivore)** is expressed using a mathematical formula. The number of prey (X) increases rapidly according to the growth rate a per unit time (aX) as long as there is grass to feed on. It is assumed that there is always plenty of grass. On the other hand, when a prey (X) and a predator (Y) meet, the former will be preyed upon with a certain probability, b (predation rate). The prey will decrease in proportion to b, X, and Y (Fig. 3.24). Since both prey and predator must exist and

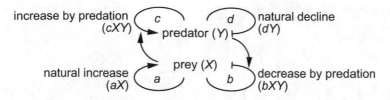

Fig. 3.24 The relationship between prey and predators

meet in order for the former to be eaten by the latter, the rate of decrease of the prey due to predation is bXY.[3]

On the other hand, predators naturally decline at a rate d (dY). However, when a prey and a predator meet, the latter preys on the former and multiplies by gaining nourishment. So, predators increase in proportion to c, X, and Y (i.e., cXY). In summary, we have the following nonlinear ordinary differential equations:

$$\frac{dX}{dt} = aX - bXY \tag{3.5.1a}$$

$$\frac{dY}{dt} = cXY - dY \tag{3.5.1b}$$

The unit time is 1 day, dt=0.01, Tmax=10000 (i.e., 100 days). The parameters are a=d=1, b=0.1, c=0.2, the initial values are X(1)=30, Y(1)=10, and the initial condition is that prey outnumber predators. Then the program will look like "script3_5A.m" below, and the result will look like Fig. 3.25.

script3_5A.m

1	`dt=0.01;Tmax1=10000;`	Or, "dt=0.002; Tmax1=500000;"
2	`X=zeros(1,Tmax1);X(1)=30;`	Number of prey, X
3	`Y=zeros(1,Tmax1);Y(1)=10;`	Number of predators, Y
4	`a=1;b=0.2;c=0.1;d=1;`	
5	`for T=1:Tmax1-1`	
6	`    X(T+1)=X(T)+dt*( a*X(T) -b*X(T)*Y(T));`	
7	`    Y(T+1)=Y(T)+dt*( c*X(T)*Y(T) -d*Y(T) );`	
8	`end`	

(continued)

---

[3]This concept is important because it can be applied to biochemical reactions. For example, let the concentrations of proteins A and B be A and B. If the interaction of A and B causes the reaction to proceed with reaction probability α, the reaction rate can be written as αAB. This is because the reaction can only occur when A and B exist and meet with each other.

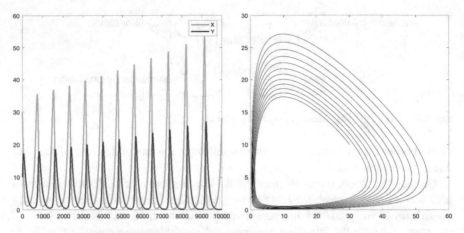

Fig. 3.25 Example of Euler method calculation results 1

9	`figure('Position',[0 300 1000 400]);`	
10	`subplot(1,2,1);plot(1:Tmax1,X,'g',1:` `Tmax1,Y,'r');legend('X','Y');`	Plot the time variation of X and Y
11	`subplot(1,2,2);plot(X, Y, 'k');`	Plot X and Y along horizontal and vertical axes

In line 10, the number of prey (X), shown by the green line, and the number of predators (Y), shown by the red line, are plotted in the left panel to show how they change over time. In the second panel, the horizontal axis is the number of prey (X) and the vertical axis is the number of predators (Y) starting from the point X=30, Y=10. When the number of predators decreases sufficiently, the number of prey increases and the number of predators begins to increase again.

In this example, we set dt=0.01, but please try to change the first line to "dt = 0.002; Tmax1 = 50000;." Although we change dt, the simulation period remains the same at 100 days because Tmax1×dt=100. In this case, we get the result shown in Fig. 3.26. Compared to Fig. 3.25, the results look quite different. In particular, the results of the X-Y plot on the right are quite different. Which result is more correct?

3.5.2 Numerical Calculation Using ode45

In Sect. 3.1.2, the numerical results using the exact solution and the Euler method are almost identical (Fig. 3.6), but the equation in Eqs. (3.5.1a) and (3.5.1b) actually has a large error in the Euler method. If the exact solution is known, such as in Eq. (3.1.2), there is no need for approximate calculation like the Euler method.

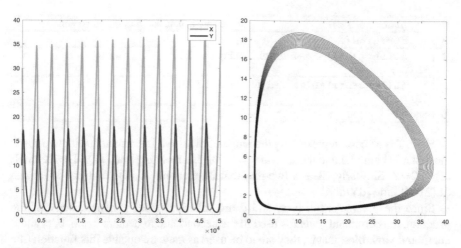

Fig. 3.26 Example of Euler method calculation results 2

However, for many nonlinear differential equations, the exact solution is not known. So, it is necessary to approximate the solution in some way.

Even with the Euler method, this error can be reduced by making dt smaller. In fact, the error in Fig. 3.26 should be smaller than that in Fig. 3.25. But the smaller dt is, the more computation time is required. In this book, we do not go into the details of numerical methods, but MATLAB provides functions for approximating ordinary differential equations with high accuracy, so let's try these functions.

In the following, we will use *ode45*, which is the most common numerical function. In order to do this, we need to prepare "a function describing an equation," "a time vector and its range," and "a matrix containing variables other than time and its initial values." But it takes some time getting used to, and the following explanation is a bit difficult, so you may skip it.

First, rewrite X as L_1 and Y as L_2 in Eqs. (3.5.1a) and (3.5.1b) and transform it as follows:

$$\frac{dL_1}{dt} = aL_1 - bL_1L_2 \tag{3.5.2a}$$

$$\frac{dL_2}{dt} = cL_1L_2 - dL_2 \tag{3.5.2b}$$

If we rewrite this equation for *ode45*, we get the following. Save this as a new function with the file name "script3_5eq.m."

script3_5eq.m	
1 `function dL=script3_5eq(T,L)`	The function name needs to be the same as its file name

(continued)

2	`global a b c d`	Declare a, b, c, and d as global variables
3	`dL=zeros(2,1);`	
4	`dL(1)=a*L(1)    -b*L(1)` `*L(2);`	Eq. (3.5.2a)
5	`dL(2)=c*L(1)*L(2)   -d*L` `(2);`	Eq. (3.5.2b)
6	`end`	

Here, T is a vector representing the range of time, and L is a matrix with two rows, the first row being L1 and the second row being L2, meaning that $L(1)=L1=X$ and $L(2)=L2=Y$. Similarly, dL is a two-row matrix, meaning $dL(1)=dL1/dt=dX/dt$ and $dL(2)=dL2/dt=dY/dt$.

Since the parameters a, b, c, and d are inherited from the main body of the program, the command *global* is used in the second line to declare that a, b, c, and d are **global variables**, that is, they are to be used as they are outside this function (the comma (,) between the variables is unnecessary here).

As explained in Sect. 2.9, variables do not need to have the same name in the body of the program and inside the function. Instead, you need to explicitly specify all the values you want to pass to the function. For example, in "script2_9A.m," line 15, we passed the information of the two players' Rock-Paper-Scissors hands to the Judge function as "`Judge(M(1,J),M(2,J))`." But here, instead of passing the four parameters a, b, c, and d, we declare them all to be global variables and treat them as common variables between the program body and the function.

The main body of the program is shown in "script3_5B.m" below, where a, b, c, and d are declared to be global variables in the first line:

script3_5B.m

1	`global a b c d`	Declare a, b, c, and d as global variables
2	`dt=0.01;Tmax1=10000;Tmax2=Tmax1*dt;`	
3	`X=zeros(1,Tmax1);X(1)=30;`	
4	`Y=zeros(1,Tmax1);Y(1)=10;`	
5	`a=1;b=0.2;c=0.1;d=1;`	
6	`[T,L]=ode45('script3_5eq', [1 Tmax2],` `[X(1);Y(1)]);`	Obtain numerical solutions of 'script3_5eq' using *ode45*
7	`figure('Position',[0 300 1000 400]);`	
8	`subplot(1,2,1);plot(T,L(:,1),'g',T,L` `(:,2),'r');`	
	`          ylim([0 40]);legend` `('X','Y');`	Plot the time variation of X (L(:,1)) and Y (L(:,2))

(continued)

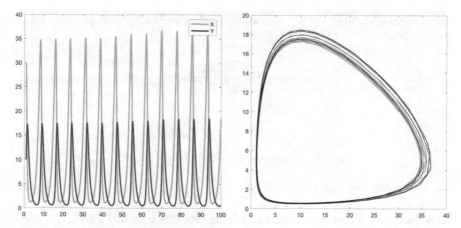

Fig. 3.27 Example of calculation results using *ode45*

9	subplot(1,2,2);plot(L(:,1),L(:,2),'k');xlim([0 40]);ylim([0 20]);	Plot X (L(:,1)) and Y (L(:,2)) along horizontal and vertical axes

In line 6, we call "script3_5eq" using *ode45*: "ode45('function name', [start time end time], [initial value of X; initial value of Y])." The first column L(:,1) is the result of X and the second column L(:,2) is the result of Y.

Note that, unlike the Euler method, the size of *dt* is not constant in *ode45*. Check the contents of T in the command window. The number of columns in T and L will not be known until the calculation is run, since the time ticks will vary as needed, but you can use the size command to find out after the run.

Now, when we run "script3_5B.m" using *ode45*, we get the result as shown in Fig. 3.27. This is quite different from Fig. 3.25, but similar to Fig. 3.26, which confirms that even the Euler method gives good results when *dt* is small.

In addition to *ode45*, MATLAB provides many other functions for numerical calculation of differential equations. However, these functions can only be applied to ordinary differential equations, that is, equations involving only the derivative of one variable such as time. When focusing on biological phenomena in multicellular organisms, it is necessary to take into account the spread in the spatial direction. In this case, we need to consider spatial differentiation as well as time differentiation, so we need to solve **partial differential equations** instead of ordinary differential equations. In this case, the Euler method or a similar method is basically used for numerical calculations, and we will use the Euler method for numerical calculations in this book.

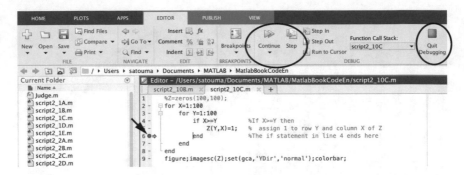

Fig. 3.28 Setting a breakpoint

Exercise 3.5.2a Solve Eqs. (3.2.1a), (3.2.1b), and (3.2.1c) using *ode45*.

Exercise 3.5.2b Solve Eqs. (3.2.2a), (3.2.2b), (3.2.2c), (3.2.2d), and (3.2.2e) using *ode45*.

Exercise 3.5.2c Solve Eqs. (3.2.3a), (3.2.3b), and (3.2.3c) using *ode45*.

Exercise 3.5.2d Solve Eqs. (3.3.1a) and (3.3.1b) using *ode45*.

Exercise 3.5.2e Solve Eqs. (3.4.1a), (3.4.1b), and (3.4.1c) using *ode45*.

Note 5. Setting Breakpoints

In most of the previous bug examples, error messages were displayed in the command window, but if no error messages are displayed, it is necessary to check how the contents of each variable change during the execution of the program. For this purpose, MATLAB allows you to set **breakpoints**.

As shown by the arrow in Fig. 3.28, when you click on the right side of the line number in the editor, a red circle appears. This is a breakpoint, and if you execute the program with this set, the program will pause just before executing the line where the breakpoint is set. In Octave, breakpoints can be set to the left of the line number.

In *Note 4*, I explained that if you disable "script2_10C.m" by adding "%" to the first line, the Z value will be wrong after "script2_10B.m" is executed. At this point, let's set a breakpoint at line 6 of "script2_10C.m" (Fig. 3.28). At first, the value of Z is calculated with X=1, Y=1, and the calculation is paused at line 6. You can find out what the value of Z is by typing Z in the command window or by running "imagesc(Z)." You will notice that the value of Z is wrong from the beginning, so the first line must be "Z=zeros (100,100);."

The green arrow just the right of the red circle indicates the line that is currently paused, and clicking on the "**Continue**" icon will restart the program

(continued)

until it stops at the breakpoint again. The "**Step**" icon will execute the program one line at a time. When you have fixed the problem, click on the "**Quit Debugging**" icon to return to the normal mode.

Chapter 4
Simulating Temporal and Spatial Changes in Biological Phenomena

4.1 Time Variation of Spatial Pattern Formation by Diffusive Materials

In the previous chapter, we discussed the mathematical model of ordinary differential equations, which considers only the time derivative without considering the spatial extension. However, when considering life phenomena in multicellular organisms, it is necessary to consider the interaction between cells laid out in space. In addition, the substances produced by each cell do not necessarily stay in that cell. Many **secreted ligands** are released and diffuse out of the cell and then bind to receptors on surrounding cells to transmit information between cells (Fig. 4.1). In this section, we will consider the phenomenon of secretory ligands diffusing in the extracellular space.

In Sect. 3.1, we considered the situation where protein E is synthesized and degraded in a single cell, but here we do not consider such synthesis and degradation. Instead, let us consider the case where E has the property of secretion and diffuses in 1D space (*x*-axis). In this case, it is known that the following relationship holds for the change in the concentration E of the diffusible substance E in the temporal and spatial directions. E is a function of time and space and is written as $E(t, x)$, but (t,x) is omitted.

$$\frac{\partial E}{\partial t} = d \frac{\partial^2 E}{\partial x^2} \tag{4.1.1}$$

Here, d is the diffusion coefficient that represents the diffusion rate of E. This equation says that "**the time variation of E is proportional to the second-order derivative of E in space.**" It is called **partial differentiation** because it involves differentiation with respect to two variables, t and x, and is distinguished from ordinary differentiation. In the case of partial differentiation, we use the partial differentiation symbol ∂ (called partial, round, del, etc.). This means that when

© The Author(s), under exclusive license to Springer Nature Singapore Pte Ltd. 2022
M. Sato, *Getting Started in Mathematical Life Sciences*, Theoretical Biology,
https://doi.org/10.1007/978-981-19-8257-6_4

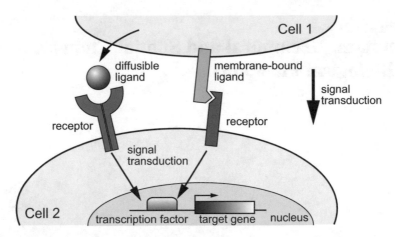

Fig. 4.1 Signal transduction through ligands and receptors

you are differentiating one variable, the other variables are considered constant. In other words, when you differentiate with respect to t, you consider x as a constant, and when you differentiate with respect to x, you consider t as a constant. Equations that include ordinary differential are called ordinary differential equations, and equations that include partial differential are called partial differential equations. However, please be assured that you do not need to be aware of the difference between ordinary and partial differentiation in programming.

4.2 Intuitive Explanation of Diffusion Equation

The reason why such an equation represents the diffusion of E can be derived strictly from the theory of statistical mechanics, but that is beyond the scope of this book, so we will try to give an intuitive explanation here. For details, please see the references [3]. For example, suppose that at a certain time t, E is distributed along the x-axis as shown in Fig. 4.2. If we focus on a point ① in space, the diffusible substance E diffuses from a place of higher concentration to a place of lower concentration in proportion to the difference in concentration (gradient). When the gradient on the right side is r_1 (white arrow) and the gradient on the left side is l_1 (black arrow), r_1 represents the inflow rate to point ① and l_1 represents the outflow rate from point ①. So, the rate of change of the amount of E at point ① is $r_1 - l_1$, taking the difference. Since point ① is located on a straight line, the gradient is the same on both the right and left sides ($r_1 = l_1$), and $r_1 - l_1$ is equal to 0, which means that the concentration of E does not change at point ①.

On the other hand, in the differential equation, $\frac{\partial E}{\partial x}$ means the gradient of E, and $\frac{\partial^2 E}{\partial x^2}$ means the gradient of the gradient of E again. At point ①, the gradient of E is simply the slope of E at point ①, but to consider the gradient of the gradient, consider the

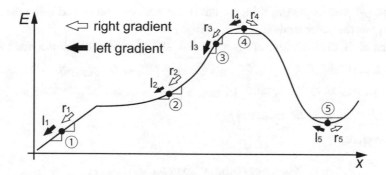

Fig. 4.2 Intuitive understanding of the diffusion equation

gradient r_1 a little to the right of ① and the gradient l_1 a little to the left of ①. The difference between r_1 and l_1, $r_1 - l_1$, is the gradient of the slope of E, or **the second-order derivative**. You are calculating the same thing as the difference between the inflow and outflow velocities that we just considered.

What about at point ②? In the same way as before, $\frac{\partial^2 E}{\partial x^2}$ represents the difference between the gradients of E just to the right and to the left of point ②, $r_2 - l_2$, which is the difference between the inflow from the right and the outflow to the left. At point ②, the gradient on the right side is steeper than the gradient on the left side, so $r_2 - l_2 > 0$ and $\frac{\partial E}{\partial t} > 0$, meaning that E is increasing with time. Conversely, at point ③, the gradient on the left side is steeper than the gradient on the right side, so the outflow to the left is faster than the inflow from the right, and E will decrease with time. Note that at point ④, the slope on the left is positive and the slope on the right is negative. In this case, the outflow is to the right and to the left. Since $r_4 < 0$ and $l_4 > 0$, $r_4 - l_4$ is negative, E will decrease with time.

Exercise 4.2 Consider what happens to E at point ⑤ in Fig. 4.2.

4.3 Calculation of the Diffusion Equation

4.3.1 Numerical Algorithm for Diffusion

In the previous chapter, we introduced the Euler method for solving ordinary differential equations. Now, how can we solve the diffusion equation? Let us assume that $d = 1$ in Eq. (4.1.1) and solve the following equation:

$$\frac{\partial E}{\partial t} = \frac{\partial^2 E}{\partial x^2}$$

(4.1.2)

The $\frac{\partial^2 E}{\partial x^2}$ part may seem difficult, but it can basically be solved using the same concept as the Euler method.

First of all, the derivative of x is $\frac{\partial E}{\partial x} = \lim\limits_{\Delta x \to 0} \frac{E(x+\Delta x) - E(x)}{\Delta x}$, and if dx is a small value, then $\frac{\partial E}{\partial x} = \frac{E(x+dx) - E(x)}{dx}$. If $F = \frac{\partial E}{\partial x}$, then $\frac{\partial^2 E}{\partial x^2} = \frac{\partial}{\partial x}\left(\frac{\partial E}{\partial x}\right) = \frac{\partial F}{\partial x}$. Since $\frac{\partial F}{\partial x} = \frac{F(x+dx) - F(x)}{dx}$, apply $F(x) = \frac{\partial E}{\partial x} = \frac{E(x+dx) - E(x)}{dx}$ to this. $F(x+dx) = \frac{E(x+2dx) - E(x+dx)}{dx}$, so $\frac{\partial F}{\partial x} = \left(\frac{E(x+2dx) - E(x+dx)}{dx} - \frac{E(x+dx) - E(x)}{dx}\right)/dx$.

Therefore,

$$\frac{\partial^2 E}{\partial x^2} = \frac{E(x+2dx) - 2E(x+dx) + E(x)}{dx^2}.$$

In addition, you learned that the definition of derivative can be considered either the gradient on the right side or the gradient on the left side of the x you are interested in. In other words,

$$\frac{\partial E}{\partial x} = \lim\limits_{\Delta x \to 0} \frac{E(x+\Delta x) - E(x)}{\Delta x} = \lim\limits_{\Delta x \to 0} \frac{E(x) - E(x-\Delta x)}{\Delta x}.$$

This of course holds for the derivative of F as well, so we can modify it a bit. Applying $F(x) = \frac{\partial E}{\partial x} = \frac{E(x+dx) - E(x)}{dx}$ to $\frac{\partial F}{\partial x} = \frac{F(x) - F(x-dx)}{dx}$, we get

$$\frac{\partial^2 E}{\partial x^2} = \frac{E(x+dx) - 2E(x) + E(x-dx)}{dx^2}.$$

As was the case with the time derivative, the computer cannot handle the notion of **extrema** (values as close to zero as possible), where dx is some small distance. We will discuss the criteria for setting the values of dx and d later (see Sect. 4.5.1), but for now, let's consider this small distance dx as the size of a cell. Of course, when considering the motion of diffusive molecules, it makes sense to think of things on a scale of thousands of the size of the cell, but for simplicity, let's think of molecules moving in units of the size of each cell, and let's assume $dx = 1$ and $d = 1$.

Note that the values of dx and d vary greatly depending on the scale of the phenomenon we are looking at. dx represents the minute length, and the diffusion coefficient d represents the diffusion rate, or the area diffused per unit time, which is proportional to the square of the unit length in a two-dimensional plane. Therefore, if $dx = 1$ μm, $d = 1$ μm^2/s, it is the same as $dx = 0.001$ mm, $d = 0.000001$ mm^2/s.

Although $dx = 1$ may not seem like a minute length, for the sake of clarity, let's consider it on a cellular scale. The cells are arranged in series, $0, 1, 2, 3, \ldots$, and the concentration of E in those cells is denoted as $E_0, E_1, E_2, E_3, \ldots$. The distance between the cells is dx. If we assume that $E(x-dx) = E_0$, $E(x) = E_1$, $E(x+dx) = E_2$, then $\frac{\partial^2 E}{\partial x^2} = \frac{E_2 + E_0 - 2E_1}{dx^2}$.

Fig. 4.3 Inflow and outflow between cells

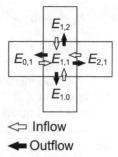

$\Leftarrow$ Inflow

$\blacktriangleleft$ Outflow

Since the diffusion equation in two dimensions is the sum of the diffusion in the x and y directions, we can write $\frac{\partial E}{\partial t} = \left(\frac{\partial^2}{\partial x^2} + \frac{\partial^2}{\partial y^2}\right)E$. Note that $\frac{\partial^2}{\partial x^2} + \frac{\partial^2}{\partial y^2}$ is written as Δ (**Laplacian**), so you can also write $\frac{\partial E}{\partial t} = \Delta E$ (in some cases, ∇^2 is used instead of Δ).

If the cells are lined up vertically and horizontally, and the horizontal position of the cells is m and the vertical position is n, and concentration of E in these cells is denoted as $E_{m,n}$ ($E_{0,1}$, $E_{1,1}$, $E_{2,1}$, $E_{1,0}$, $E_{1,2}$) (Fig. 4.3), then $\frac{\partial^2 E}{\partial x^2} = \frac{E_{2,1} + E_{0,1} - 2E_{1,1}}{dx^2}$ and $\frac{\partial^2 E}{\partial y^2} = \frac{E_{1,2} + E_{1,0} - 2E_{1,1}}{dy^2}$. If we assume that the cell is a square and $dx = dy$, then

$$\left(\frac{\partial^2}{\partial x^2} + \frac{\partial^2}{\partial y^2}\right)E = \frac{E_{2,1} + E_{0,1} + E_{1,2} + E_{1,0} - 4E_{1,1}}{dx^2}.$$

If we add the diffusion coefficient d to this, we get

$$\left(\frac{\partial^2}{\partial x^2} + \frac{\partial^2}{\partial y^2}\right)E = d\frac{E_{2,1} + E_{0,1} + E_{1,2} + E_{1,0} - 4E_{1,1}}{dx^2}.$$

This means that the change in E due to diffusion is proportional to the amount obtained by adding the values of E at the top, bottom, left, and right and subtracting four times the value of E in the middle (Fig. 4.3).

Let's consider again why the equation derived in this way expresses diffusion. When we consider the time variation of $E_{1,1}$ in the center cell in Fig. 4.3, the amount of inflow from the left to the center (white arrow) is proportional to $E_{0,1}$, and the amount of outflow from the center to the left (black arrow) is proportional to $E_{1,1}$, so the net change is $d(E_{0,1} - E_{1,1})$. The same is true for the right, up, and down. If we consider the same thing for the interaction with the cells on the right, top, and bottom, $d(E_{0,1} - E_{1,1}) + d(E_{2,1} - E_{1,1}) + d(E_{1,2} - E_{1,1}) + d(E_{1,0} - E_{1,1}) = d(E_{2,1} + E_{0,1} + E_{1,2} + E_{1,0} - 4E_{1,1})$, and the change in E due to diffusion is proportional to the sum of the values of E in the top, bottom, left, and right, minus four times the value of E in the middle. If we consider the time derivative of the left-hand side of Eq. (4.1.1), that is, $\frac{\partial E}{\partial t} = d\left(\frac{\partial^2}{\partial x^2} + \frac{\partial^2}{\partial y^2}\right)E$, according to the Euler method,

$$E_{1,1}(t+dt) = \frac{d(E_{2,1}(t) + E_{0,1}(t) + E_{1,2}(t) + E_{1,0}(t) - 4E_{1,1}(t))dt}{dx^2}$$
$$+ E_{1,1}(t) \tag{4.1.3}$$

Now, if we know the spatial distribution of E at time t, $E(t)$, we can calculate the distribution of E at time $t + dt$ as $E(t + dt)$.

Since E is a function of space and time, or time t and spatial coordinates x and y in a two-dimensional plane, it is formally written as $E(t, x, y)$. In MATLAB, this E is represented by a matrix. For example, it is a three-dimensional matrix with X=1: 100, Y=1:100, and T=1:Tmax. As explained in Sects. 2.10.1 and 2.11.2, when we consider displaying it in *imagesc*, we can easily handle it by making the first dimension the Y-axis (Y row), the second dimension the X-axis (X column), and the third dimension the T-axis (Fig. 2.31). Therefore, the matrix representing $E(t, x, y)$ will be denoted as "E(Y,X,T)" in MATLAB.

As in the previous chapter, the numbers Y, X, and T must be positive integers, and t is replaced by T and $t + dt$ by T+1. $E_{1,1}(t)$ is replaced by E(Y,X,T), $E_{2,1}(t)$, $E_{0,1}(t)$, $E_{1,2}(t)$, $E_{1,0}(t)$ become E(Y,X+1,T), E(Y,X-1,T), E(Y+1,X,T) and E(Y-1,X,T), respectively, so Eq. (4.1.3) can be expressed in MATLAB style as

```
E(Y,X,T+1)=d(E(Y,X+1,T)+E(Y,X-1,T)+E(Y+1,X,T)+E(Y-1,
X,T)-4*E(Y,X,T))*dt/dx2 + E(Y,X,T)
```

Now you can imagine how to calculate this in a real program, can't you?

4.3.2 Calculating Diffusion Using for Statement

In reality, time T varies from 1 to Tmax(=100), and spatial coordinates X and Y vary from 1 to 100, so let's use a *for* statement loop to calculate this one by one. Also, the initial distribution of E at time T=1 must be determined in advance. As shown in Fig. 4.4, if we set E=1 only for the rectangular region of X=41:60, Y=41:60, and E=0 for the other regions, we get the following:

script4_3A.m

1	Xmax=100; Tmax=100;	
2	dt=0.1;d=1;dx=1;	
3	E=zeros(Xmax,Xmax,Tmax);	
4	E(41:60,41:60,1)=1;	Initial distribution of E
5	for T=1:Tmax	
6	for X=1:Xmax	
7	for Y=1:Xmax	

(continued)

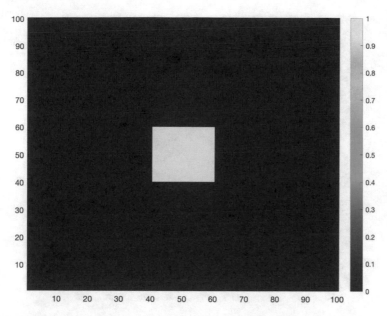

Fig. 4.4 Initial distribution of E

8	E(Y,X,T+1)=dt*(d/dx/dx*(E(Y,X+1,T)+E (Y,X-1,T)	
	+E(Y+1,X,T)+E(Y-1,X,T)-4*E(Y,X,T)))+E (Y,X,T);	Diffusion of E by the Euler method
9	end	
10	end	
11	end	
12	for T=1:Tmax	
13	imagesc(E(:,:,T),[0 1]);axis xy;colorbar;	
14	pause(0.001);	
15	end	

The third line "E=zeros(Xmax, Xmax, Tmax);" defines a three-dimensional matrix E with Xmax in the X direction, Xmax in the Y direction, and Tmax in the time T direction. The fourth line "E(41:60, 41:60, 1)=1;" gives the initial condition that E=1 in the square region of X=41:60, Y=41:60 at time T=1 (Fig. 4.4). The Eulerian calculation on the eighth line is printed on a new line because it is long. But it should be entered on a single line without line breaks. If it is too long to read horizontally, you may add three periods (...) at the end of the line. The next line after the three periods will be considered the same line.

Lines 12 to 15 show the animation of the result obtained by combining the "for T=1:Tmax" loop with *imagesc* and *pause*. Line 13 shows the pattern of E at time T using *imagesc*. Since E is a 3D matrix, we consider it as a 2D matrix by fixing the

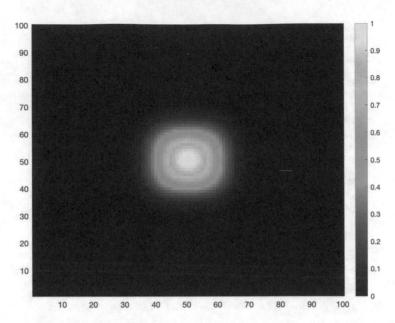

Fig. 4.5 Distribution of E after diffusion

time to T as "E(:,:,T)" (Fig. 2.31). Additionally, in the instruction "imagesc (E(:,:,T),[0 1]);", [0 1] specifies that E should be displayed in the range 0 to 1 (see Sect. 2.11.2, "script2_11C"). This is to prevent *imagesc* from changing the color display range on its own when E diffuses and the value of E becomes smaller.

However, when you run such a code, you will get the following error message:

Index in position 2 is invalid. Array indices must be positive integers or logical values

Error in script4_3Adiffusion (line 8)

E(Y,X,T+1)=dt*(d/dx/dx*(E(Y,X+1,T)+E(Y,X-1,T)+E(Y+1, X,T)+E(Y-1,X,T)-4*E(Y,X,T)))+E(Y,X,T);

This is because the problem occurs when X or Y is 1 or 100. For example, when X=1, X-1 in "E(Y, X-1, T)" will be 0, and when X=100, X+1 in "E(Y,X+1, T)" will be 101. However, since the second dimension of E (X-axis) is only defined from 1 to 100, X=0 and X=101 do not exist, and an error message will appear.

We have no choice but to modify line 6 to be "for X=2:Xmax-1" and line 7 to be "for Y=2:Xmax-1" so that X and Y do not exceed the range of 1 to 100. When you run the program, you will get the result that the distribution of E is diffused as shown in Fig. 4.5.

However, since we have limited the scope of the *for* statement to X=2:99 and Y=2:99, we have not calculated the boundary region of the 100×100 2D region. We would like to calculate the boundary region somehow, but how should we think about diffusion in the boundary region?

Fig. 4.6 Diffusion in the
boundary region

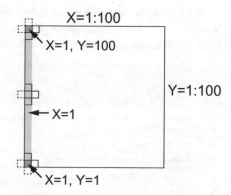

The above problem arises from the fact that no boundary is considered for the diffusion equation. Here, the boundary is assumed to be a wall, and a non-flux boundary condition is introduced assuming no external inflow from or outflow to the outside of the 100×100 region. There are other possibilities, such as **periodic boundary conditions** where the top, bottom, left, and right sides of the region are connected to form a circle, but for now, let's think that "the boundary has walls."

Here, the idea that diffusion is determined by the inflow from neighboring cells to the central cell minus the outflow to neighboring cells is important (Fig. 4.3). For example, for the left end of the region, that is, Y=2:99 in the X=1 column, we do not need to consider the interaction with the cells on the left side, but only the inflow and outflow from the right, upper, and lower cells (Fig. 4.6). So, the calculation of diffusion is

```
E(Y,X,T+1)=d(E(Y,X+1,T)+E(Y+1,X,T)+E(Y-1,X,T)-3*E(Y,
X,T))*dt/dx2+E(Y,X,T);
```

Note that since we are not considering the outflow to the left, we have "`-3*E(Y, X,T)`" instead of "`-4*E(Y,X,T)`."

Also, for X=1; Y=1; in the lower left corner, we only need to consider the inflow and outflow between the upper and right cells, so

```
E(Y,X,T+1)=d(E(Y,X+1,T)+E(Y+1,X,T)-2*E(Y,X,T))*dt/
dx2+E(Y,X,T);
```

Since we are not considering the outflow to the left and down, the outflow term is "`-2*E(Y,X,T)`." Putting all this together in MATLAB code, we have:

```
X=1;
for Y=2:99
    E(Y,X,T+1)=d(E(Y,X+1,T)+E(Y+1,X,T)+E(Y-1,X,T)-3*E(Y,
X,T))*dt/dx2+E(Y,X,T);
end
X=1;Y=1;
E(Y,X,T+1)=d(E(Y,X+1,T)+E(Y+1,X,T)-2*E(Y,X,T))*dt/
dx2+E(Y,X,T);
```

We can do the same for the right, top, and bottom boundary regions, such as X=100, Y=1, Y=100, etc. Now you can calculate the diffusion for all the boundary regions. You can complete the code "script4_3B.m" by yourself.

script4_3B.m

1	`Xmax=100;Tmax=100;`	
2	`dt=0.1;d=1;dx=1;`	
3	`E=zeros(Xmax,Xmax,Tmax);`	
4	`E(41:60,41:60,1)=1;`	
5	`tic;`	Start time measurement
6	`for T=1:Tmax`	
7	`  for X=2:Xmax-1`	Calculate the region of X=2:99, Y=2:99
8	`    for Y=2:Xmax-1`	
9	`      E(Y,X,T+1)=dt*(d/dx/dx*(E(Y,X+1,T)+E(Y,X-1,T)`	
	`        +E(Y+1,X,T)+E(Y-1,X,T)-4*E(Y,X,T)))+E(Y,X,T);`	
10	`    end`	
11	`  end`	
12	`  for X=2:Xmax-1`	Calculate the region of X=2:99
13	`    Y=1;`	Calculate the region of Y=1
14	`    E(Y,X,T+1)=dt*(d/dx/dx*(E(Y,X+1,T)+E(Y,X-1,T)`	
	`      +E(Y+1,X,T)-3*E(Y,X,T)))+E(Y,X,T);`	
15	`    Y=Xmax;`	Calculate the region of Y=100
16	`    E(Y,X,T+1)=dt*(d/dx/dx*(E(Y,X+1,T)+E(Y,X-1,T)`	
	`      +E(Y-1,X,T)-3*E(Y,X,T)))+E(Y,X,T);`	
17	`  end`	
18	`  for Y=2:Xmax-1`	Calculate the region of Y=2:99
19	`    X=1;`	Calculate the region of X=1
20	`    E(Y,X,T+1)=dt*(d/dx/dx*(E(Y,X+1,T)+E(Y+1,X,T)`	
	`      +E(Y-1,X,T)-3*E(Y,X,T)))+E(Y,X,T);`	
21	`    X=Xmax;`	Calculate the region of X=100

(continued)

22	`E(Y,X,T+1)=dt*(d/dx/dx*(E(Y,X-1,T)+E(Y` `+1,X,T)`	
	`+E(Y-1,X,T)-3*E(Y,X,T)))+E(Y,X,T);`	
23	`end`	
24	`X=1; Y=1;`	Calculate the region of X=1, Y=1
25	`E(Y,X,T+1)=dt*(d/dx/dx*(E(Y,X+1,T)+E(Y` `+1,X,T)-2*E(Y,X,T)))+E(Y,X,T);`	
26	`X=Xmax; Y=1;`	Calculate the region of X=100, Y=1
27	`E(Y,X,T+1)=dt*(d/dx/dx*(E(Y,X-1,T)+E(Y` `+1,X,T)-2*E(Y,X,T)))+E(Y,X,T);`	
28	`X=1; Y=Xmax;`	Calculate the region of X=1, Y=100
29	`E(Y,X,T+1)=dt*(d/dx/dx*(E(Y,X+1,T)+E` `(Y-1,X,T)-2*E(Y,X,T)))+E(Y,X,T);`	
30	`X=Xmax; Y=Xmax;`	Calculate the region of X=100, Y=100
31	`E(Y,X,T+1)=dt*(d/dx/dx*(E(Y,X-1,T)+E` `(Y-1,X,T)-2*E(Y,X,T)))+E(Y,X,T);`	
32	`end`	
33	`toc;`	End time measurement
34	`for T=1:Tmax`	
35	`imagesc(E(:,:,T),[0 1]);axis xy;colorbar;`	
36	`pause(0.001);`	
37	`end`	

When you run it, you will get the same result as in Fig. 4.5. In addition, the command window will display the calculation time as "The elapsed time is ~ seconds." This is due to "tic;" in line 5 and "toc;" in line 33. *tic* starts the stopwatch, *toc* stops it, and the elapsed time is displayed in the command window. The reason we are doing this is that there are several ways to calculate diffusion, and it is better to use the faster method. Since lines 34–37 are just to show the result in animation, we only measure the time required for the calculations in lines 6–32.

4.3.3 Calculating Diffusion Using del2

Although "script4_3B.m" is a rather complicated code, there is actually a command *del2* in MATLAB that calculates the diffusion. If E is a two-dimensional matrix, it holds "**del2(E,dx)=ΔE/4**." Since E is actually a three-dimensional matrix, we can regard it as a two-dimensional matrix by fixing the time to T as "E(:,:,T)," just as we did for *imagesc* (Fig. 2.31). Then the diffusion equation of E (Eq. (4.1.1)) is expressed as follows:

E(:,:,T+1)=4*d*del2(E(:,:,T),dx)*dt+E(:,:,T);

It is only one line. If dx=1, you can omit dx (such as "del2(E(:,:,T))").
Rewrite "script4_3B.m" using *del2*. "script4_3C.m" should look as follows:

script4_3C.m

1	Xmax=100;Tmax=100;	
2	dt=0.1;d=1;dx=1;	
3	E=zeros(Xmax,Xmax,Tmax);	
4	E(41:60,41:60,1)=1;	Or, "E(1:20, 41:60, 1)=rand(20);"
5	tic;	
6	for T=1:Tmax-1	
7	E(:,:,T+1)=4*dt*d*del2(E(:,:,T))+E (:,:,T);	Calculation of the diffusion of E using *del2*
8	end	
9	toc;	
10	for T=1:Tmax	
11	imagesc(E(:,:,T),[0 1]);axis xy; colorbar;	
12	pause(0.001);	
13	end	

It's much simpler now. By the way, the calculation speed is also faster with *del2*.
On my PC, it took 0.076263 seconds for "script4_3B.m" and 0.022585 seconds for
"script4_3C.m," more than three times faster. The calculations here are simple, so in
either way, you will get the result at once. However, in more complex cases, it may
take more than an hour to complete the calculation. In such cases, being several
times faster is important.

You may think that since *del2* is simpler and faster, you can always use *del2* to
calculate the diffusion. But be careful because *del2* has a problem in calculating the
boundary. So far, the initial distribution of E has been "E(41:60,41:60,1)
=1;," but try changing the fourth line of "script4_3C.m" to "E(1:20,41:60,1)
=rand(20);." The initial conditions will be as shown in Fig. 4.7. The diffusible
protein E is present at the boundary of the region and its concentration is random. If
you start the calculation from such initial conditions, you will end up with a messy
pattern as shown in Fig. 4.8. As you can see, *del2* sometimes has problems in
calculating the boundary region.

4.3.4 Fast Calculation of Diffusion Using Matrices

In fact, there is a computation algorithm that can handle the boundary region
computation without any problem, is faster than *del2*, and is simpler than

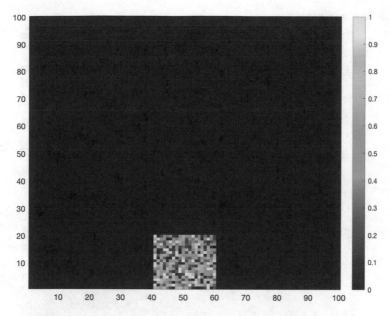

Fig. 4.7   Random initial distribution

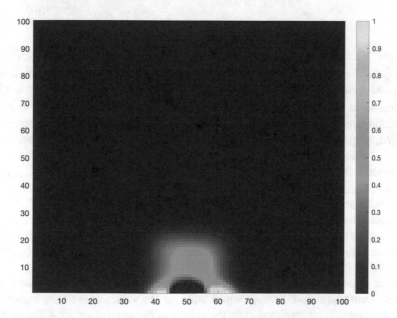

Fig. 4.8   Calculation result using *del2*

Fig. 4.9 Shifting matrices

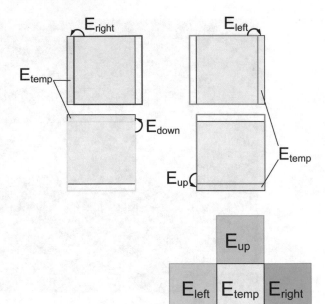

Fig. 4.10 Four adjacent cells

"script4_3B.m." As you can see in Fig. 4.3, calculating the diffusion is the same as adding the top, bottom, left, and right values, subtracting the middle value times four, and dividing by $dx2$, or $\frac{E_{2,1}+E_{0,1}+E_{1,2}+E_{1,0}-4E_{1,1}}{dx^2}$. In "script4_3B.m," we used *for* statements to calculate one cell at a time, but since MATLAB is good at matrix calculations, we can use matrices to do the equivalent calculations all at once.

First, prepare the matrices E_{right}, E_{left}, E_{up}, E_{down} by shifting the original matrix E (called E_{temp} here to make it easier to distinguish from others) in the direction of up, down, left, and right. Fig. 4.9 shows the positions of the matrices E_{temp}, E_{right}, E_{left}, E_{up}, E_{down}, and Fig. 4.10 shows the positions of the neighboring cells. E_{right} has information about the position of one cell to the right of E_{temp}. In MATLAB, this can be written as "Eright(:,1:99)=Etemp(:,2:100);." However, for the rightmost cell in the region, that is, X=100, there are no cells to the right of it, so the same information for X=100 is inserted. In other words, "Eright(:,100) =Etemp(:,100);."

In the same way, we can define Eleft, Eup, and Edown and then calculate **Eright +Eleft+Etop+Edown-4*Etemp**, which will sum the values of E above, below, left, and right at each cell level, multiply the values in the same place by 4, and subtract them. You will see that the algorithm is very simple. The code will look like "script4_3D.m" below:

script4_3D.m

1	`Xmax=100;Tmax=100;`	
2	`dt=0.1;d=1;dx=1;dx2=dx*dx;`	
3	`E=zeros(Xmax,Xmax,Tmax);`	
4	`E(41:60,41:60,1)=1;`	Or, "E(1:20, 41:60, 1)=rand (20);"
5	`Etemp=zeros(Xmax,Xmax);Eright=zeros (Xmax,Xmax);`	
	`Eleft=zeros(Xmax,Xmax);Eup=zeros (Xmax,Xmax);`	
	`Edown=zeros(Xmax,Xmax);`	Prepare the matrices Etemp, Eright, Eleft, Eup, Edown
6	`tic;`	
7	`for T=1:Tmax-1`	
8	`  Etemp=E(:,:,T);`	Calculation of the diffusion of E using the matirices
9	`  Eright(:,100)=Etemp(:,100);Eright (:,1:99)=Etemp(:,2:100);`	
10	`  Eleft(:,1)=Etemp(:,1);Eleft(:,2:100) =Etemp(:,1:99);`	
11	`  Eup(100,:)=Etemp(100,:);Eup(1:99,:) =Etemp(2:100,:);`	
12	`  Edown(1,:)=Etemp(1,:);Edown(2:100,:) =Etemp(1:99,:);`	
13	`  E(:,:,T+1)=dt*(d*(Eright+Eleft+Eup +Edown-4*Etemp)/dx2)+Etemp;`	
14	`end`	
15	`toc;`	
16	`for T=1:Tmax`	
17	`  imagesc(E(:,:,T),[0 1]);axis xy; colorbar;`	
18	`  pause(0.001);`	
19	`end`	

Line 5 defines that Etemp, Eright, Eleft, Eup, Edown are matrices of size 100×100 (Xmax-by-Xmax), but these matrix definitions are not absolutely necessary. The important point is that Etemp, Eright, Eleft, Eup, and Edown are all 2D matrices lacking the third dimension along the time axis, T. Line 8 "Etemp=E (: , : , T) ;" assigns the spatial information (X- and Y-axis information) of E at time T to Etemp. In line 9, the information in column 100 of Etemp is assigned to column 100 of Eright, and the information in columns 2–100 of Etemp is assigned to columns 1–99 of Eright. Lines 10–12 create Eleft, Eup, and Edown from Etemp in the same way. In line 13, the Euler method is used to calculate "E (: , : , T+1) " at time T+1.

The computation time is 0.011461 s, which is about twice as fast as that of "script4_3C.m" using *del2*. It is about six times faster than that of "script4_3B.m" using the *for* statement, which means that the computation takes 10 min instead of 1 h. (Octave does not seem to be as fast.) Furthermore, check that the calculation does not fail even if the fourth line is set to "E(1:20, 41:60, 1)=rand(20);" using the initial condition in Fig. 4.7.

Since the code for calculating the diffusion is complicated and hard to read, we can use *function* we learned in Sect. 2.9 to create a new function for calculating the diffusion. Then you can do a fast calculation of diffusion in the same way as *del2*.

Exercise 4.3.4 Make a function to calculate diffusion ("Diffusion.m") using the algorithm shown in Sect. 4.3.4 and modify "script4_3D.m" using it.

Note 6. Common Bugs #4: Wrong Algorithm

In some cases, the program may behave strangely even though there is no problem superficially. In such cases, there is usually no error message. In this case, the construction of the logic, that is, the algorithm, is considered to be problematic. You need to think about the role of each instruction in the program and debug it by using breakpoints effectively. And often, such programs are grammatically correct, but contain improper positioning of certain instructions.

In "script3_4D.m," the initial values of I and S are calculated in lines 10–11, and the differential equations of the SIR model are calculated using these values in lines 12–16. Actually, it is very important that "I(1)=N(J) *0.01;" in line 10 and "S(1)=N(J)−I(1);" in line 11 are in this position, but beginners may often make a mistake on this point.

The initial values of I and S depend on the value of N. In the *for* statement loop from line 9–19, the value of J is varied and the vector N is used as N(J). If "N(J)=N(J)−I(1);" is placed before line 9, the initial values of I and S cannot be set according to the value of N(J). In such a case, you can set a breakpoint at line 19, for example, and check how the values of S, I, and R actually change with J.

4.4 Pattern Formation by Morphogens

Our body is made up of many cells, which communicate with each other through a variety of molecules. Such cell-to-cell communication controls various properties of cells [4, 6]. Figure 4.1 shows a schematic representation of the mechanism of information transfer between two cells. Although there are many possible mechanisms for information transfer from the upper cell to the lower cell, we will consider the most common example via **secretory ligands**. Secretory ligands (proteins in most cases) are produced in the upper cells, secreted out of the cells, and diffuse outside the cells (therefore, they are also called **diffusible ligands**). On the other hand, the lower cells, which are the recipients of the information, have receptors for the ligands. When the ligand binds to the receptor, it triggers the activation of **signal transduction** from the receptor to the nucleus.

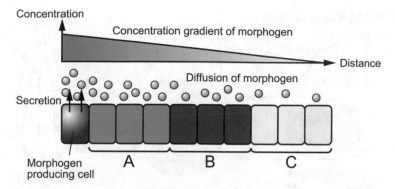

Fig. 4.11 Positional information in a tissue

Signal transduction is a chain reaction of many molecules, but in this book, I would like to give a simplified conceptual explanation. In many cases, signaling ultimately regulates the activity of **transcription factors**, which in turn regulate the **transcription** of **target genes**. Transcription factors bind to **DNA sequences** in the vicinity of the target gene and thereby regulate the transcription of the target gene, and in many cases, the activity of the transcription factor is regulated by signal transduction.

Thus, cell-to-cell **signal transduction** is established by controlling the transcription of **target genes** through **ligand**, **receptors**, **signal transduction**, and **transcription factors**. In other cases, membrane-bound ligands are used instead of secretory diffusible ligands, and information is transmitted only between neighboring cells (Fig. 4.1).

In the case of diffusible ligand-mediated signal transduction, the concentration of the ligand plays an important role. In Fig. 4.11, the leftmost cell produces the secretory ligand, which creates a concentration gradient with higher concentration on the left side and lower concentration on the right side. The surrounding cells gradually differentiate into different cells according to this concentration. Here, the cells in the high concentration area differentiate into A cells, those in the medium concentration area into B cells, and those in the low concentration area into C cells. **Morphogens** are molecules that control the differentiation of cells according to their concentration and thus control the spatial pattern of the entire tissue. When morphogens are produced by some cells in a tissue and diffuse outside the cell, a concentration gradient of morphogens is formed, and the differentiation of surrounding cells is controlled according to this concentration [4].

The mechanism of morphogen action has been reported in many studies, and it is known to be very complex. Here we adopt the simplest approach and assume that morphogens simply diffuse out of the cell.

Many molecules are known to act as morphogens, but the *Drosophila* (fruit fly) **Decapentaplegic (Dpp)** protein is very famous and is known to regulate patterning along the anterior–posterior axis in the formation of fly wings [10, 11]. The adult wing primordium, a disk-shaped tissue found in the larval stage, grows during the

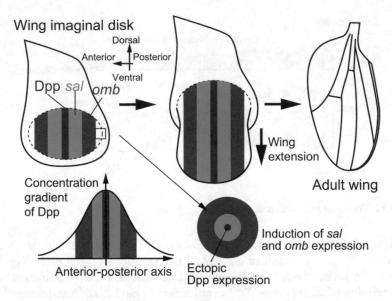

Fig. 4.12 Wing patterning by Dpp

pupal stage to form the adult wing (Fig. 4.12). Wing veins and sensory organs are regularly arranged in the adult wing, and it is thought that the information that forms these spatial patterns, especially the positional information in the anterior–posterior direction, is given by the concentration gradient of Dpp.

In the adult wing primordium, Dpp is produced in the striped region shown in blue, which creates a concentration gradient along the anterior–posterior axis (Fig. 4.12). Corresponding to this Dpp concentration gradient, the *sal* and *omb* genes are expressed in the surrounding cells. In Fig. 4.12, *sal* is expressed in the region where the concentration of Dpp is high (green) and *omb* is expressed in the region where the concentration of Dpp is low (red), parallel to the stripe where Dpp is produced. The names of proteins such as Dpp are written in regular font, and the names of genes such as *sal* and *omb* are written in italics.

The expression of *sal* and *omb* disappears in mutants of Dpp and Dpp receptor, but how can we confirm that the expression of these genes is actually regulated in a Dpp concentration-dependent manner? In the upper left corner of Fig. 4.12, we can see that *sal* and *omb* are not expressed in the area far from the Dpp expression area. Here, a small number of cells ectopically expressing Dpp were artificially generated in this area in the form of a spot. If this Dpp diffuses, it will form concentric gradients, and *sal* and *omb* should also be expressed in concentric circles. In fact, as shown in the lower right corner of Fig. 4.12, the cells that express Dpp show a concentric expression pattern, with cells closer to the center expressing *sal* and cells farther away expressing *omb*, indicating that Dpp does indeed act as a morphogen [10, 11].

4.5 Pattern Formation in Fly Wing

4.5.1 Calculating the Diffusion of Dpp

In Sect. 4.3, we worked on the calculation of diffusion, and pattern formation by morphogens is an excellent topic to apply the calculation of diffusion. In this section, we will study the effect of Dpp on wing patterning in the fly, and although we would like to use D as the variable for the concentration of Dpp, we will avoid D since we will be using another gene, *Dll*, later. Since Dpp is an evolutionarily conserved protein and is a homolog of vertebrate BMP4, let B be the variable representing the concentration of Dpp. If we set the diffusion coefficient of B as d, the degradation rate coefficient of B as k, and the production rate of B as c, we get the following equation:

$$\frac{\partial B}{\partial t} = d\Delta B - kB + c \qquad (4.5.1)$$

The right-hand side means that B diffuses according to the diffusion coefficient d, degrades at a rate of k per unit time, and is produced at a rate of c. The term $d\Delta B$ represents diffusion, and $-kB + c$ represents the chemical reaction. Thus, the equation including diffusion and chemical reaction is called the **reaction–diffusion equation**. For details of mathematical models using the reaction–diffusion equation, please refer to other books, for example, [1].

Let us assume $dt = 0.1$, $dx = 1$, $T_{max} = 500$, $d = 1$, $k = 0.1$. c is the production rate of B, which is the distribution pattern shown in Fig. 4.13 and contains spatial information. Let "c=zeros(Xmax,Xmax); c(:,48:53)=0.5;" so that B is produced at a rate of 0.5 per unit time in the central striped region (X=48:53). Since it is very difficult to measure the diffusion coefficient of Dpp in tissues and the amount of Dpp produced per unit time, and there is no such measurement data, we assume that the values of these parameters are arbitrary.

If you use *del2* to calculate the diffusion, the code will look like "script4_5A.m" below. If your PC is not powerful enough or if you are using Octave, it may take too long to display the animation, so you may use "for T=1:20:Tmax" in line 8 to display the image every 20 frames.

script4_5A.m	

1	`Xmax=100;Tmax=500;`	
2	`B=zeros(Xmax,Xmax,Tmax);`	
3	`dt=0.1; dx=1;d=1;k=0.1;`	Or, "d=10;"
4	`c=zeros(Xmax,Xmax);c(:,48:53)=0.5;`	Spatial pattern of c, the production rate of B

(continued)

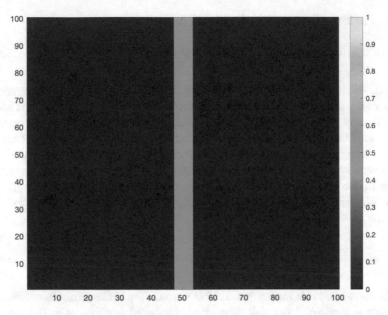

Fig. 4.13 Spatial distribution of matrix c

5	`for T=1:Tmax-1`	Reaction–diffusion calculation using *del2*
6	`B(:,:,T+1)=dt*(4*d*del2(B(:,:,T))-k*B` `(:,:,T)+c)+B(:,:,T);`	
7	`end`	
8	`for T=1:Tmax`	Or, "for T=1:20:Tmax"
9	`imagesc(B(:,:,T),[0 1]);axis xy;` `colorbar;`	
10	`pause(0.0001);`	
11	`end`	

By transforming Eq. (4.5.1) according to the Euler method, you can see that "B
(:,:,T+1)=dt*(4*d*del2(B(:,:,T))-k*B(:,:,T)+c)+B(:,:,
T);" in line 6. Make sure that *dt* is multiplied by the entire right-hand side of
Eq. (4.5.1).

You will get the result shown in Fig. 4.14. Compared to Fig. 4.13, you can see
that B diffuses around the produced region. To plot the concentration of B at Y=51
(dashed line in Fig. 4.14) on the vertical-axis, type "plot(B(51,:,Tmax));" in
the command window (Fig. 4.15). In "script4_5A.m," "[0 1]" in the *imagesc*
instruction in line 9 sets the range of values as 0 to 1. Even if the value is greater
than 1, it will be displayed as 1. If you set this to "[0 3]" or simply "imagesc(B
(:,:,T));," the result will be displayed correctly even if the value is greater
than 1.

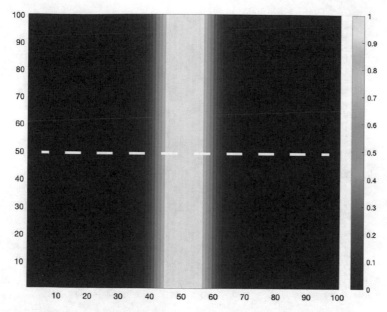

Fig. 4.14 Spatial distribution of matrix B

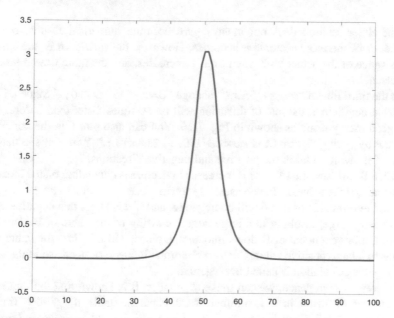

Fig. 4.15 Spatial distribution of B at Y=51

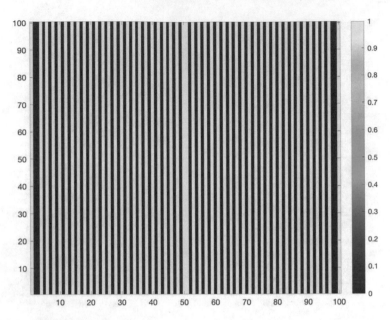

Fig. 4.16 Divergent calculation result

The Euler method does not always give accurate numerical results, but the smaller dt is, the more accurate it becomes. However, the smaller dt is, the smaller the advance of the actual time when time T increases, and the more time it takes to calculate.

In the third line of "script4_5A.m," change "d=1;" to "d=10;." Since d is the diffusion coefficient, the rate of diffusion will be 10 times faster than before. You will get a messy result as shown in Fig. 4.16. You can also plot it in the command window by typing "plot(1:Xmax,B(51,:,Tmax));." You will see that the value of B swings a lot in the positive and negative directions.

When B diffuses, the value of B increases or decreases depending on the location. In this case, if the value of B decreases, "4*d*del2(B(:,:,T))-k*B(:,:,T) +c" on the right side of line 6 will be negative, and if d is large, this negative value will be very large, resulting in a large negative swing of the value of B. The next moment, a large amount of B flows into such a place, and this time the value of B swings to a large positive value. Such an abnormal increase or decrease in the value of computer calculation is called **divergence**.

To prevent divergence, we can make dt smaller. It is known that the numerical solution for diffusion in a two-dimensional plane is stable if $\frac{d\,dt}{dx^2} < \frac{1}{4}$ (in one dimension, $\frac{d\,dt}{dx^2} < \frac{1}{2}$). $dt = 0.1$, $dx = 1$, $d = 1$ satisfies this condition, but $d = 10$ does not. We can make dt small enough to satisfy this condition by setting "dt=0.01;dx=1;d=10;k=0.1;" in the third line to avoid divergence, which should result in a smooth calculation similar to Fig. 4.14. Also, dx should be a small value, but if $dx = 0.1$, then d and dt should be small enough. In reality, there are other terms besides diffusion that may affect the results, but we can use this conditional formula as a guide to adjust the values of *dt, dx*, and *d* [2, 3]. The "implicit scheme" is a known method for removing such restrictions on *dt* and *dx*, but it is beyond the level of this book, so please see [3].

Exercise 4.5.1 Modify "script4_5A.m" using the diffusion function generated in Exercise 4.3.4.

4.5.2   *Expression of Target Genes of Dpp*

How can we express the expression of the target gene of Dpp (B) in an equation? Let the expression level of *sal*, the target gene of B, be the variable *S*. If the degradation rate of *S* per unit time is the same as that of *B* and *k*, and the induction rate of *S* by *B* (the rate of transcription) is *a*, the differential equation of *S* is as follows:

$$\frac{\partial S}{\partial t} = aB - kS \qquad (4.5.2)$$

Assuming that $d = 1$, $k = 0.1$, $a = 1$, and Eqs. (4.5.1) and (4.5.2) are calculated simultaneously, the code will look like the following:

script4_5B.m

1	`Xmax=100;Tmax=500;`	
2	`B=zeros(Xmax,Xmax,Tmax);S=zeros(Xmax,Xmax,Tmax);`	
3	`dt=0.1; dx=1;d=1;k=0.1;a=1;`	
4	`c=zeros(100,100);c(:,48:53)=0.5;`	
5	`for T=1:Tmax-1`	
6	`B(:,:,T+1)=dt*(4*d*del2(B(:,:,T))-k*B(:,:,T)+c)+B(:,:,T);`	
7	`S(:,:,T+1)=dt*(a*B(:,:,T)-k*S(:,:,T))+S(:,:,T);`	Production and degradation of S
8	`end`	
9	`figure('Position',[0 300 1000 400]);`	
10	`for T=1:Tmax`	Or, "for T=1:20:Tmax"
11	`subplot(1,2,1);imagesc(B(:,:,T),[0 1]);axis xy;colorbar;`	
12	`subplot(1,2,2);imagesc(S(:,:,T),[0 1]);axis xy;colorbar;`	
13	`pause(0.0001);`	
14	`end`	

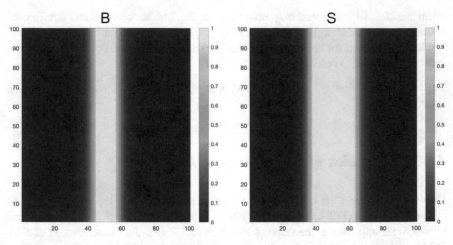

Fig. 4.17 Spatial distribution of the matrix B

You will get the result as shown in Fig. 4.17. The left panel shows B and the right panel shows S. In line 9, a horizontal window with a width of 1000 pixels and a height of 400 pixels is prepared, and "subplot(1,2,1);" in line 11 indicates to draw in the first (i.e., left) panel of the two horizontal panels, while "subplot (1,2,2);" in line 12 indicates to draw in the second (i.e., right) panel. As you can see, you are able to simulate the induction of *sal* expression according to the morphogen concentration gradient of Dpp.

Now, as you have seen in "script4_3C.m," the calculation using *del2* may have some problems in handling the boundary region. On the other hand, the matrix-based method shown in "script4_3D.m" solves the boundary problem and is faster than *del2*.

Exercise 4.5.2a Replace the diffusion calculation in "script4_5B.m" with the faster version (see "script4_3D.m"). An example is shown below:

script4_5C.m

1	Xmax=100;Tmax=500;	
2	B=zeros(Xmax,Xmax,Tmax);S=zeros(Xmax, Xmax,Tmax);	
3	dt=0.1;dx=1;dx2=dx*dx;d=1;k=0.1;a=1;	
4	c=zeros(100,100);c(:,48:53)=0.5;	
5	Btemp=B(:,:,1);Bright=Btemp; Bleft=Btemp;Bup=Btemp;Bdown=Btemp;	
6	for T=1:Tmax-1	
7	Btemp=B(:,:,T);	Fast calculation of the diffusion of B using matrices

(continued)

8	`Bright(:,Xmax)=Btemp(:,Xmax);Bright(:,1:Xmax-1)=Btemp(:,2:Xmax);`	
9	`Bleft(:,1)=Btemp(:,1);Bleft(:,2:Xmax)=Btemp(:,1:Xmax-1);`	
10	`Bup(Xmax,:)=Btemp(Xmax,:);Bup(1:Xmax-1,:)=Btemp(2:Xmax,:);`	
11	`Bdown(1,:)=Btemp(1,:);Bdown(2:Xmax,:)=Btemp(1:Xmax-1,:);`	
12	`B(:,:,T+1)=dt*(d/dx2*(Bright+Bleft+Bup+Bdown-4*Btemp)-k*Btemp+c)+Btemp;`	
13	`S(:,:,T+1)=dt*(a*Btemp-k*S(:,:,T))+S(:,:,T);`	
14	`end`	
15	`figure('Position',[0 300 1000 400]);`	
16	`for T=1:Tmax`	Or, "for T=1:20:Tmax"
17	`subplot(1,2,1);imagesc(B(:,:,T),[0 1]);axis xy;colorbar;`	
18	`subplot(1,2,2);imagesc(S(:,:,T),[0 1]);axis xy;colorbar;`	
19	`pause(0.0001);`	
20	`end`	

Exercise 4.5.2b Modify "script4_5C.m" using the diffusion function generated in Exercise 4.3.4.

4.6 Pattern Formation in Fly Leg

4.6.1 Regulation of Dll expression by Dpp and Wg

The wing patterning discussed in the previous section is a simple story that target gene *sal* is expressed in a region parallel to the Dpp stripes, and I think you can understand it intuitively. However, this is not enough for you to appreciate the importance of building a mathematical model and conducting simulations. Next, I would like to consider an example that is a little more complicated and difficult to understand intuitively.

Like wings, the legs of flies are formed by the growth of a disk-shaped adult leg primordium during the pupal stage (Fig. 4.18). At this time, the leg primordium is compartmentalized into concentric circles, and the cells closer to the center elongate to the tip of the leg and the cells farther from the center contribute to the base of the leg. The nature of the leg tip is determined by the concentric expression of a gene called ***Distalless (Dll)*** at the center of the leg primordium (Fig. 4.19). But how is this concentric expression pattern determined? In this process, both Dpp and a diffusible ligand called **Wingless (Wg)** act as morphogens to regulate the expression of *Dll*

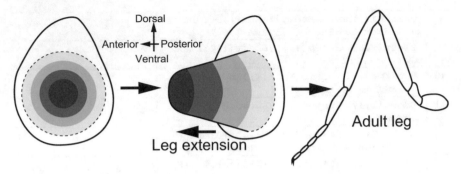

Fig. 4.18 Growth of the leg primordium

Fig. 4.19 Expression
domains of Dll, Dpp,
and Wg

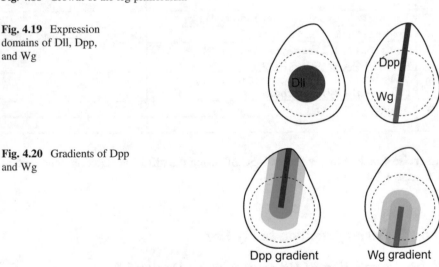

Fig. 4.20 Gradients of Dpp
and Wg

(Fig. 4.19) [10–13]. Both are produced in a striped region in the middle of the leg primordium, but Dpp is expressed dorsally and Wg is expressed only ventrally, both of which regulate *Dll* expression. As shown in Fig. 4.20, Dpp and Wg are thought to form a concentration gradient in the dorsal and ventral regions, respectively.

In the paper by Lecuit and Cohen, it was stated that "concentric expression regions are formed when *Dll* is transcribed in the presence of both Dpp and Wg inputs" [13], but I think this is not enough to convince us whether this is really the case or not. Can you easily imagine that *Dll* expression domain becomes concentric when its transcription is activated by Dpp and Wg? Let us consider this issue by constructing a mathematical model of the reaction–diffusion of Dpp and Wg and the expression of the target gene, *Dll*.

Let the concentration of Dpp be B, the concentration of Wg be W (as Wg belongs to a family of secretory proteins called Wnt), the concentration of *Dll* be D, the diffusion coefficient of B and W be d, and the degradation rate of B, W, and D be k. If

Fig. 4.21 Signal
transduction downstream of
Dpp and Wg

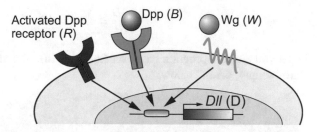

the production rates of B and W are c_b and c_w, respectively, the differential equations for B and W are as follows. The equations for B and W are exactly the same as in the previous section:

$$\frac{\partial B}{\partial t} = d\Delta B - kB + c_b \tag{4.6.1a}$$

$$\frac{\partial W}{\partial t} = d\Delta W - kW + c_w \tag{4.6.1b}$$

$$\frac{\partial D}{\partial t} = ? - kD \tag{4.6.1c}$$

How should we think about the differential equation for D (Eq. (4.6.1c))? It is obvious that the degradation of D is $-kD$. But how should we think about the production of D under the control of B and W? When Dpp and Wg bind to their receptors, they regulate the transcription of the Dll gene in the nucleus through their respective signaling pathways (Fig. 4.21).

If we simply assume that B and W independently induce the production of D, then $\frac{\partial D}{\partial t} = aB + bW - kD$ (a and b are arbitrary coefficients), but this obviously does not work. In other words, Dll is induced in the region shown on the left in Fig. 4.20. Similarly, Dll is induced in the region shown on the right in Fig. 4.20, where Wg signaling is activated by bW, and the Dll pattern will not be concentric as in Fig. 4.19.

The important point here is that "Dll is transcribed when there are **both** Dpp and Wg inputs." In other words, B and W do not work independently; the value of D needs to increase only when both signals are present. This may sound difficult, but think about what kind of operation it would be if D does not increase for any value of W when $B = 0$, and likewise when $W = 0$. Of course, you can't add them together. In this case, you have to think of **multiplication**: considering the product of B and W, if one of them is zero, the result is zero, and conversely, if both of them are large, the result is also large.[1] In this case, Dll is transcribed in proportion to the concentration of both Dpp and Wg, and if the coefficient is a, the equation is as follows:

---

[1] This is the same logic used in the Lotka–Volterra equation in Sect. 3.5.1. When a prey (X) and a predator (Y) meet, the latter is assumed to increase in proportion to the density of both and the coefficient c (cXY). The same is true for the SIR model in Sect. 3.4.1.

$$\frac{\partial D}{\partial t} = aBW - kD \tag{4.6.2}$$

If you create a program based on Eqs. (4.6.1a), (4.6.1b) and (4.6.2), it will look like "script4_6A.m" below:

---

script4_6A.m

1	`Xmax=100;Tmax=500;`	
2	`B=zeros(Xmax,Xmax,Tmax);W=B;D=B;`	
3	`dt=0.1;d=1;k=0.1;a=1;`	
4	`cb=zeros(Xmax,Xmax);cb(51:100,41:60)=0.5;`	
5	`cw=zeros(Xmax,Xmax);cw(1:50,41:60)=0.5;`	
6	`for T=1:Tmax-1`	
7	`  B(:,:,T+1)=dt*(4*d*del2(B(:,:,T))-k*B(:,:,T)+cb)+B(:,:,T);`	
8	`  W(:,:,T+1)=dt*(4*d*del2(W(:,:,T))-k*W(:,:,T)+cw)+W(:,:,T);`	
9	`  D(:,:,T+1)=dt*(a*B(:,:,T).*W(:,:,T)-k*D(:,:,T))+D(:,:,T);`	Transcription and degradation of D
10	`end`	
11	`figure('Position',[0 400 1000 250]);`	
12	`for T=1:Tmax`	Or, "for T=1:20:Tmax"
13	`  subplot(1,3,1);imagesc(B(:,:,T),[0 1]);axis xy;colorbar;`	
14	`  subplot(1,3,2);imagesc(W(:,:,T),[0 1]);axis xy;colorbar;`	
15	`  subplot(1,3,3);imagesc(D(:,:,T),[0 1]);axis xy;colorbar;`	
16	`pause(0.0001);`	
17	`end`	

The most important thing to note here is that in line 9, aBW is written as "a*B(:,:,T).*W(:,:,T)." Since BW needs to be the Hadamard product of the matrices B and W, we need to add a period prior to "*" to make it ".*" (Fig. 1.15). Since a is a scalar variable, you don't need to add a period after "a." The result is shown in Fig. 4.22. The left panel shows B, the center shows W, and the right panel shows D.

4.6.2 Effects of Activated Dpp Receptors

In the paper by Lecuit and Cohen, there is another interesting experimental result [13]. Normally, Dpp receptors are activated only when Dpp binds to them, but it is

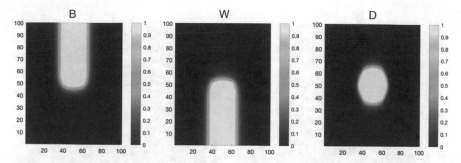

Fig. 4.22 Calculation results of B, W, and D

Fig. 4.23 Induction of Dll
by activated Dpp receptor

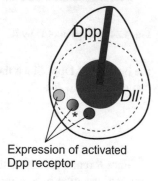

Expression of activated
Dpp receptor

possible to artificially create activated receptors that are always active even when
Dpp is not present (Fig. 4.21). Also, by using a method called **mosaic analysis**, it is
possible to express the activated receptor in a mosaic fashion in a small portion of
cells in a tissue. How does the expression of *Dll* change in this case? If Dpp signaling
alone, independent of Wg signaling, can induce *Dll* expression, then *Dll* expression
would be upregulated in any cell that expresses the activated Dpp receptor. However, as shown in Fig. 4.23, the expression of *Dll* is induced only in the vicinity of
the Wg expression region [13]. In addition, there is a gradient in the ectopic
expression of *Dll*, as shown in Fig. 4.23.

How can we incorporate such a mosaic expression of the activated Dpp receptor
into a mathematical model? So far, we have not considered the variables that
represent the receptors of Dpp and Wg and have assumed that the expression of
Dll is immediately induced when the ligand is present. Of course, we could introduce
a new variable to represent the activity of Dpp and Wg receptors, but in this case,
let's simplify it and represent the spatial distribution of activated Dpp receptors by
R (Fig. 4.21). Since the diffusion, degradation, and production of Dpp and Wg are
unrelated to the activation of Dpp receptors, there is no need to change the differential equations for B and W in Eqs. (4.6.1a), (4.6.1b), and (4.6.1c). How about
Eq. 4.6.2? *Dll* transcription is regulated according to the concentration of Dpp and
Wg. In this case, the value of B represents the activation of the Dpp receptor due to

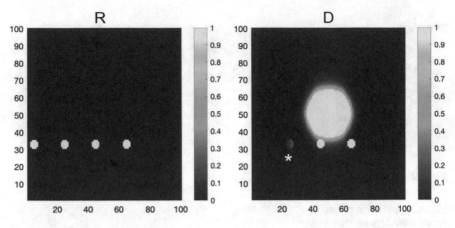

Fig. 4.24 Induction of D by R

the binding of Dpp. Since the effect of the activated Dpp receptor is added to B, we can modify B to $B + R$.

$$\frac{\partial D}{\partial t} = a(B + R)W - kD \qquad (4.6.3)$$

Since R represents the activated Dpp receptor, we don't need the time information and can just give it as a two-dimensional matrix "R=zeros(Xmax, Xmax);." However, since the value is zero in all regions, we need to add a positive value such as 1 in the mosaic region expressing the activated Dpp receptor. Let us first prepare a 5-by-5 matrix, "Mosaic=ones(5,5);" and assign it to matrix R. For example, if we set "R(48:52,48:52)=Mosaic;" then only the center of matrix R will have a value of 1, and the Dpp receptor will always be active in this region. Using this method, we can make the Dpp receptor (R) active in the region where Y=31:35 and X=3:7, 23:27, 43:47, 63:67 (Fig. 4.24). You can write a code similar to "script4_6B.m" below:

script4_6B.m

1	Xmax=100;Tmax=500;	
2	B=zeros(Xmax,Xmax,Tmax);W=B;D=B; R=zeros(Xmax,Xmax);	
3	dt=0.1;d=1;k=0.1;a=1;	
4	cb=zeros(Xmax,Xmax);cb(51:100,41:60) =0.5;	
5	cw=zeros(Xmax,Xmax);cw(1:50,41:60)=0.5;	
6	Mosaic=ones(5,5);	

(continued)

7	`R(31:35,3:7)=Mosaic;R(31:35,23:27)` `=Mosaic;`	
8	`R(31:35,43:47)=Mosaic;R(31:35,63:67)` `=Mosaic;`	Specify the distribution of activated receptor R
9	`for T=1:Tmax`	
10	`  B(:,:,T+1)=dt*(4*d*del2(B(:,:,T))-k*B` `(:,:,T)+cb)+B(:,:,T);`	
11	`  W(:,:,T+1)=dt*(4*d*del2(W(:,:,T))-k*W` `(:,:,T)+cw)+W(:,:,T);`	
12	`  D(:,:,T+1)=dt*(a*(B(:,:,T)+R).*W(:,:,` `T)-k*D(:,:,T))+D(:,:,T);`	Calculate the production and degradation of D
13	`end`	
14	`figure('Position',[0 400 1000 250]);`	
15	`for T=1:20:Tmax`	
16	`  subplot(1,3,1);imagesc(B(:,:,T),[0 1]);` `axis xy;colorbar;`	
17	`  subplot(1,3,2);imagesc(W(:,:,T),[0 1]);` `axis xy;colorbar;`	
18	`  subplot(1,3,3);imagesc(D(:,:,T),[0 1]);` `axis xy;colorbar;`	
19	`  pause(0.001);`	
20	`end`	

When this is done, the result shown in Fig. 4.24 is obtained (B and W are omitted since they are the same as in Fig. 4.22). In fact, the shape of the population of cells expressing the activated Dpp receptor is not square but round, so in the example, line 6 is "`Mosaic=[0 1 1 0; 1 1 1 1; 1 1 1 1; 1 1 1 1; 0 1 1 0];`" to make it look round, but this is not essentially important. Either way, we can see the gradient of *Dll* expression in the simulation, as indicated by the asterisks in Figs. 4.23 and 4.24.

Exercise 4.6.2a   Replace the diffusion calculation in "script4_6B.m" with the faster version (see "script4_3D.m").

Exercise 4.6.2b   Modify "script4_6B.m" using the diffusion function generated in Exercise 4.3.4.

Note 7. Common Bugs #5: Mathematical Problems

So far, bugs have been programming problems, but there are also cases where bugs are caused by mathematical problems. For example, if the transformation of equations according to the Euler method (Sect. 3.1.1) or the calculation of the Hadamard product (Sect. 1.8.2) is not done correctly, the result will be completely wrong. For the Euler method in particular, you should write the equation on a piece of paper, transform the equation, and then convert it to MATLAB code. Errors can be caused by a simple typo, such as if you

(continued)

Note 7. Common Bugs #5: Mathematical Problems (continued)
misplace the parentheses or forget to add a period (.), which means Hadamard product.

For example, in "script3_3A.m," line 8, if

"G(T+1)=G(T)+dt*(-k1*G(T)*(1+a*I(T)) + d(T));" is

miswritten as

"G(T+1)=G(T)+dt*(-k1*G(T)*(1+a*I(T) + d(T)));,"

there will be no error message. But the result will be completely different from Fig. 3.12. If you do not notice this bug, you will overlook the error in the calculation result. As long as there are no error messages, it is important to be aware of how this mathematical model behaves. If the blood glucose level G, which should normally change with food intake, does not change at all, you have to realize that there is something wrong with the program, that there is a problem in the calculation of G.

4.7   Turing Model

4.7.1   Reaction–Diffusion Equation Consisting of Activators and Inhibitors

In the mathematical models that we have been working on so far, we have substituted real numerical values or physical quantities for variables, such as the concentration of molecules or the number of organisms. This makes it possible to determine the values of parameters and variables based on actual measurements and, conversely, to experimentally verify predictions obtained by simulation.

On the other hand, in the life sciences, there are some values that are very difficult to measure or quantities that cannot be clearly defined in physical terms. For example, the physical entity of the diffusible factor discussed in this section is not clear, and the physical entity of the differentiation state of cells discussed in Sect. 4.9 is also not clear. In such a case, we can construct a "**phenomenological mathematical model**" that focuses on the observed phenomenon itself, without being bound by the actual physical quantity. Phenomenological mathematical models are not necessarily based on real molecular mechanisms but play an important role in inferring and demonstrating the mechanisms behind the phenomena based on the results obtained.

Many people think of Alan Turing's **Turing model** when they hear of mathematical models of biological pattern formation [1, 14]. At the time of Turing, the double-helix structure of DNA had not yet been elucidated, and the entities of morphogens, signal transduction, and transcriptional regulation discussed in the previous section were completely unknown. Nevertheless, Turing devised a mathematical model based on the reaction–diffusion equation, which consists of the diffusion of two diffusible substances and a chemical reaction, and showed that

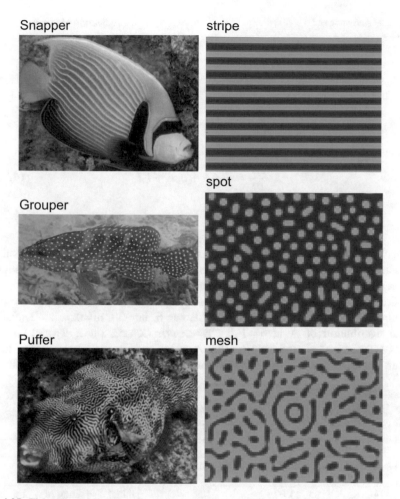

Fig. 4.25 Fish skin and Turing patterns

this model can be used to describe the patterns on the body surface of various animals, such as the patterns on the body surface of fish (Fig. 4.25).[2] Dr. Shigeru Kondo in Osaka University, Japan, has shown that the Turing model actually explains the pattern changes in the stripes of the snapper and is famous for a series of studies that integrate the theory of the Turing model with experiments [15–21].

The Turing model considers two diffusible substances, **activator** (A) and **inhibitor** (I) (Fig. 4.26). The concentrations of both are A and I, and the diffusion coefficients are d_a and d_i, respectively. If the control of I by A is a_iA (>0), the control of A by I is i_aI (<0), the control of A by A is a_aA (>0), and the control of I by I is i_iI (<0), the following differential equation is obtained:

---

[2] Photos by the author

Fig. 4.26 Activator and inhibitor

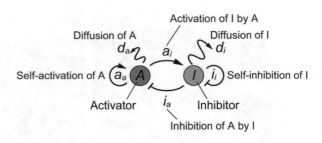

$$\frac{\partial A}{\partial t} = d_a \Delta A + a_a A + i_a I \qquad\qquad (4.7.1\text{a})$$

$$\frac{\partial I}{\partial t} = d_i \Delta I + a_i A + i_i I \qquad\qquad (4.7.1\text{b})$$

In the Turing model, it is important that the diffusion of I is faster than the diffusion of A, that is, $d_i > d_a$. But before that, we would like to consider a slightly different phenomenon using the same equations. Let us first consider the case where the diffusion of A is faster than that of I, that is, $d_a > d_i$, where Tmax=500, Xmax=100, dt=0.01, and da=20, di=0.1, aa=6, ia=-30, ai=0.5, ii=-2, and the initial distribution of A at T=1 is "A=zeros(Xmax,Xmax,Tmax);A(48: 52,48:52,1)=1;" and suppose that A exists only in the 5-by-5 region in the center. The calculation of the diffusion can be done with *del2*, which looks like:

---

script4_7A.m

1	`Tmax=500;Xmax=100;`	
2	`A=zeros(Xmax,Xmax,Tmax);I=zeros` `(Xmax,Xmax,Tmax);`	
3	`A(48:52,48:52,1)=1;`	Initial distibution of A at time T=1
4	`dt=0.01;da=20;aa=6;di=0.1;ii=-2;` `ia=-30;ai=0.5;`	
5	`for T=1:Tmax`	Calculating the reaction–difffusion of A and I using *del2*
6	`  A(:,:,T+1)=dt*(4*da*del2(A(:,:,T))` `+aa*A(:,:,T)+ia*I(:,:,T))+A(:,:,T);`	
7	`  I(:,:,T+1)=dt*(4*di*del2(I(:,:,T))` `+ii*I(:,:,T)+ai*A(:,:,T))I(:,:,T);`	
8	`end`	
9	`figure('Position',[0 300 1000 400]);`	
10	`for T=1:20:Tmax`	
11	`  subplot(1,2,1);imagesc(A(:,:,T),` `[0 2]);axis xy;colorbar;`	

(continued)

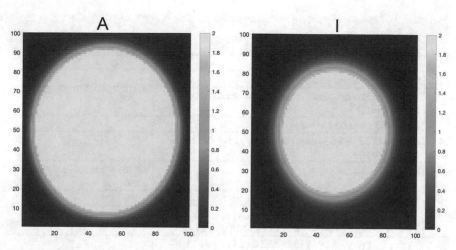

Fig. 4.27 Diffusion of A and I

12	`subplot(1,2,2);imagesc(I(:,:,T),` `[0 2]);axis xy;colorbar;`
13	`pause(0.0001);`
14	`end`

If you understand the code in the previous section, there is nothing difficult about it. When this is done, A spreads outward from the center, and a little later, I also spreads outward from the center (Fig. 4.27). One would expect that the increase in I in the center would suppress A. However, this is not the case, because the concentration of A is so high that the suppression by I is not sufficient.

4.7.2 Setting an Upper Bound on Concentration

In a real cell, the concentration of a chemical substance such as a protein cannot be infinitely high. Since there must be some upper limit, let's set the upper limits of A and I as $A_{max} = 2$ and $I_{max} = 2$, respectively, and make sure that these values do not exceed the upper limit. For example, we could use the logistic equation described in Sect. 3.2.3, but for the sake of simplicity, let's simply add the conditions $A \leq A_{max}$ and $I \leq I_{max}$ to Eqs. (4.7.1a) and (4.7.1b). To add this condition to the code, we can use the *for* and *if* statements to check if the value of A is less than A_{max} in each cell, and if it is, set the value to A_{max} and do the same for I. The code will look like the following:

script4_7B.m

1	`Tmax=500;Xmax=100;`	
2	`A=zeros(Xmax,Xmax,Tmax);I=zeros(Xmax,Xmax,Tmax);`	
3	`A(48:52,48:52,1)=1;`	
4	`dt=0.01;da=20;aa=6;di=0.1;ii=-2;ia=-30;ai=0.5;`	
5	`Amax=2;Imax=2;`	Set the upper limits Amax and Imax
6	`for T=1:Tmax`	
7	`  A(:,:,T+1)=dt*(4*da*del2(A(:,:,T))+aa*A(:,:,T)+ia*I(:,:,T))+A(:,:,T);`	
8	`  I(:,:,T+1)=dt*(4*di*del2(I(:,:,T))+ii*I(:,:,T)+ai*A(:,:,T))+I(:,:,T);`	
9	`  for Y=1:100`	Incrementing Y from 1 to 100
10	`    for X=1:100`	Incrementing X from 1 to 100
11	`      if A(Y,X,T+1)<0`	Setting the upper and lower limits of A(Y,X,T+1)
12	`        A(Y,X,T+1)=0;`	
13	`      elseif A(Y,X,T+1)>Amax`	
14	`        A(Y,X,T+1)=Amax;`	
15	`      end`	
16	`      if I(Y,X,T+1)<0`	Setting the upper and lower limits of I(Y,X,T+1)
17	`        I(Y,X,T+1)=0;`	
18	`      elseif I(Y,X,T+1)>Imax`	
19	`        I(Y,X,T+1)=Imax;`	
20	`      end`	
21	`    end`	
22	`  end`	
23	`end`	
24	`figure('Position',[ 0 300 1000 400]);`	
25	`for T=1:20:Tmax`	
26	`  subplot(1,2,1);imagesc(A(:,:,T),[0 2]);axis xy;colorbar;`	
27	`  subplot(1,2,2);imagesc(I(:,:,T),[0 2]);axis xy;colorbar;`	
28	`  pause(0.0001);`	
29	`end`	

Lines 9–22 use a double *for* statement loop to change X and Y by 1, and lines 11 to 20 use an *if* statement to make sure that A is less than or equal to Amax and I is less than or equal to Imax. Also, since A and I represent the concentration of a substance, they cannot be negative. Therefore, if the value is less than zero, we set it

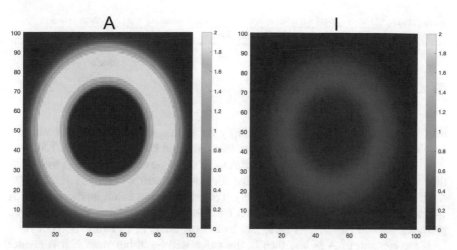

Fig. 4.28 Ripple formation

to zero. As shown in Fig. 4.28, A is suppressed in the center, and the ripples form a pattern that spreads outward.

4.7.3 Setting an Upper Bound on Concentration Without Using **if** Statement

However, the *for* statement loop in lines 9–22 is redundant, and the execution speed is slow because it changes X and Y by one. Is there a more concise and faster way? Since MATLAB is good at matrix computation, it would be concise and fast if you could do the equivalent using only matrix computation without using *for* and *if* statements. In Sect. 2.12.2, we used the property that if you make a conditional judgment using inequality or equality, it will return 1 if the condition is correct and 0 otherwise. The same method can be used for matrices.

Try typing "B=[0 1 2; 1 2 3; 2 3 4];" followed by "B>Amax" in the command window. The result will be as follows:

```
0 0 0
0 0 1
0 1 1
```

Here, a matrix has been generated in which each element is 1 if the corresponding value of matrix B is greater than Amax, and 0 otherwise. If you want the value to be Amax when it is greater than Amax, you can multiply it by Amax. So, you can use "Amax.*(B>Amax)." If the condition "B>Amax" is satisfied, the value of the corresponding element will be Amax.

Similarly, if you type "B>0," you will get 1 if the corresponding element of B is greater than 0, and 0 otherwise. So, if you want to have the original values of B when

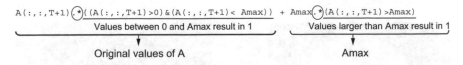

Fig. 4.29 Setting the upper bound of a matrix

they are greater than 0, you can multiply by B to get "B.*(B>0)." Note that multiplication here is the Hadamard product indicated by ".*" (see Sect. 1.8.2).

In this case, we want the elements of A to be the same as the original A as long as they are between 0 and Amax, so we first consider the matrix "(A(:,:,T+1)>0) &(A(:,:,T+1)<Amax)" (Fig. 4.29). To get the same value as the original A when it is 1, just multiply by A. Don't forget the period, since it is Hadamard product, "A(:,:,T+1).*((A(:,:,T+1)>0)&(A(:,:,T+1)<Amax))."

If the element of A is less than 0, the value will be 0, but what if it is greater than Amax? In "A(:,:,T+1)>Amax," the value will be 1 if the element of A is greater than or equal to Amax. But in this case, we want the value to be Amax, so we can multiply by Amax (Fig. 4.29). Thus, we can have "Amax.*(A(:,:,T+1)>Amax)." The value of "(A(:,:,T+1)>0)&(A(:,:,T+1)<Amax)" will result in zero if A is greater than or equal to Amax, so we can add these two terms to get

"A(:,::,T+1).*((A(:,:,T+1)>0)&(A(:,:,T+1)<Amax))+Amax.*(A(:,:,T+1)>Amax)"

If you rewrite lines 9 to 22 of "script4_7B.m," you will get the following "script4_7C.m" (Fig. 4.29). You should get the same result as "script4_7B.m" (Fig. 4.28).

script4_7C.m	

1	Tmax=500;Xmax=100;	
2	A=zeros(Xmax,Xmax,Tmax);I=zeros(Xmax,Xmax,Tmax);	
3	A(48:52,48:52,1)=1;	
4	dt=0.01;da=20;aa=6;di=0.1;ii=-2;ia=-30;ai=0.5;	
5	Amax=2;Imax=2;	
6	for T=1:Tmax	
7	A(:,:,T+1)=dt*(4*da*del2(A(:,:,T))+aa*A(:,:,T)+ia*I(:,:,T))+A(:,:,T);	
8	I(:,:,T+1)=dt*(4*di*del2(I(:,:,T))+ii*I(:,:,T)+ai*A(:,:,T))+I(:,:,T);	
9	A(:,:,T+1)=A(:,:,T+1).*((A(:,:,T+1)>0)&(A(:,:,T+1)<Amax))+Amax.*(A(:,:,T+1)>Amax);	Setting the upper and lower limits of A

(continued)

10	`I(:,:,T+1)=I(:,:,T+1).*((I(:,:,T+1)>0)&(I(:,:,T+1)<Imax))+Imax.*(I(:,:,T+1)>Imax);`	Setting the upper and lower limits of I
11	`end`	
12	`figure('Position',[ 0 300 1000 400]);`	
13	`for T=1:20:Tmax`	
14	`  subplot(1,2,1);imagesc(A(:,:,T),[0 2]);` `axis xy;colorbar;`	
15	`  subplot(1,2,2);imagesc(I(:,:,T),[0 2]);` `axis xy;colorbar;`	
16	`  pause(0.0001);`	
17	`end`	

Exercise 4.7.3 Modify "script4_7C.m" so that the ripples are formed continuously.

In "script4_7C.m," A was first produced in the center and then spread outward. If A is always produced in the center, the ripples will always be generated and will spread outward one after another. If the production of A is c, Eqs. (4.7.1a) and (4.7.1b) can be modified as follows:

$$\frac{\partial A}{\partial t} = d_a \Delta A + a_a A + i_a I + c \tag{4.7.2a}$$

$$\frac{\partial I}{\partial t} = d_i \Delta I + a_i A + i_i I \tag{4.7.2b}$$

If the initial value of A is 0, and A is produced only in the central region, c can be defined as "`c=zeros(Xmax,Xmax);c(48:52,48:52)=1;`." Then, all you have to do is to add the term "`+c`" to line 7 of "script4_7C.m." Modify "script4_7C. m," and complete "script4_7C2.m."

4.7.4 Turing Pattern

Now we are ready to proceed to the programming of the Turing model. First of all, it is important that the inhibitor I diffuses faster than the activator A, so we change the parameters accordingly. In order to set an upper limit on the rate of production rather than the concentration of A and I, we have prepared A_p and I_p as variables representing the amount of each produced, and their upper limits, A_{pmax} and I_{pmax}, respectively. There are various other ways to give the reaction terms such as A_p and I_p, but we will use the simplest method here. For details, please refer to references [17–20, 22].

$$\frac{\partial A}{\partial t} = d_a \Delta A - k_a A + A_p \;\; \left(0 \le A_p = c_1 A + c_2 I + c_3 \le A_{pmax}\right) \tag{4.7.3a}$$

$$\frac{\partial I}{\partial t} = d_i \Delta I - k_i I + I_p \;\; \left(0 \le I_p = c_4 A + c_5 I + c_6 \le I_{pmax}\right) \tag{4.7.3b}$$

Here, dt=0.1, dx=1, da=0.02, di=0.5, ka=0.03, ki=0.06, c1=0.08, c2=-0.08, c3=0.1, c4=0.11, c5=0, c6=-0.15, Apmax=0.2, Ipmax=0.5. The initial conditions for A and I are "A=zeros(Xmax,Xmax,Tmax);I=zeros(Xmax,Xmax, Tmax);A(46:50,46:50,1)=1;" Let's write the program as "script4_7D.m."

As for A and I, they are 3D matrices because we need to calculate their values at each time T and display them later in the animation. On the other hand, for A_p and I_p, we only need to temporally calculate the amount of production, so there is no need to keep the values for a long time. Ap and Ip are prepared in line 4, but it is not absolutely necessary.

If your PC is not powerful enough or if you are using Octave, the calculation may take too long. If so, please change line 1 to "Tmax=5000;Xmax=20;" and line 3 to "A(10:11,10:11,1)=1;."

script4_7D.m

1	Tmax=5000;Xmax=100;	
2	A=zeros(Xmax,Xmax,Tmax);I=zeros(Xmax, Xmax,Tmax);	
3	A(46:50,46:50,1)=1;	Or, "A(:,:,1)=rand (Xmax,Xmax);"
4	Ap=zeros(Xmax,Xmax);Ip=zeros(Xmax,Xmax);	Prepare 2D matrices Ap and Ip
5	dt=0.1;dx=1;	
6	da=0.02;di=0.5;ka=0.03;ki=0.06;	
7	c1=0.08;c2=-0.08;c3=0.1;c4=0.11;c5=0; c6=-0.15;	
8	Apmax=0.2;Ipmax=0.5;	
9	for T=1:Tmax-1	
10	Ap=c1*A(:,:,T)+c2*I(:,:,T)+c3;	Calculate the production of A as Ap
11	Ip=c4*A(:,:,T)+c5*I(:,:,T)+c6;	Calculate the production of I as Ip
12	Ap=((Ap<Apmax)&(Ap>0)).*Ap+(Ap>Apmax). *Apmax;	Set the upper and lower limits of Ap
13	Ip=((Ip<Ipmax)&(Ip>0)).*Ip+(Ip>Ipmax). *Ipmax;	Set the upper and lower limits of Ip

(continued)

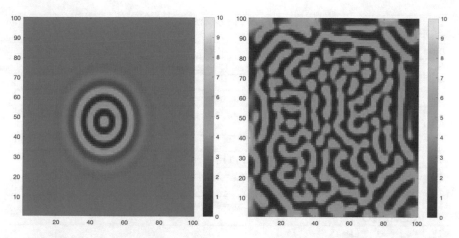

Fig. 4.30 Turing pattern using *del2*

14	`A(:,:,T+1)=dt*(4*da*del2(A(:,:,T))+Ap-ka*A(:,:,T))+A(:,:,T);`	
15	`I(:,:,T+1)=dt*(4*di*del2(I(:,:,T))+Ip-ki*I(:,:,T))+I(:,:,T);`	
16	`end`	
17	`figure('Position',[0 300 1000 400]);`	
18	`for T=1:100:Tmax`	
19	`subplot(1,2,1);imagesc(A(:,:,T),[0 10]);axis xy;colorbar;`	
20	`subplot(1,2,2);imagesc(I(:,:,T),[0 10]);axis xy;colorbar;`	
21	`pause(0.0001);`	
22	`end`	

You can see the ripples spreading outward as shown in the left side of Fig. 4.30. However, unlike Fig. 4.28, the concentric circle pattern in the center remains. In the following figures, only the value of A is shown.

Next, let's rewrite the third line as "`A(:,:,1)=rand(Xmax,Xmax);`" and start with a random initial pattern. You will see a stripe pattern like the one in the right side of Fig. 4.30. However, in this case, the pattern in the boundary region is unnaturally disordered. Similarly to Sect. 4.3.3, *del2* has problems in computing the boundary region, so follow the method in Sect. 4.3.4 and replace it with a fast calculation using matrices.

script4_7E.m

1	`Tmax=10000;Xmax=100;`	Or, "Tmax=5000;Xmax=20;

(continued)

2	`A=zeros(Xmax,Xmax,Tmax);I=zeros(Xmax,Xmax,Tmax);`	
3	`A(:,:,1)=rand(Xmax,Xmax);`	
4	`Ap=zeros(Xmax,Xmax);Ip=zeros(Xmax,Xmax);`	
5	`dt=0.1;dx=1;dx2=dx*dx;`	
6	`da=0.02;di=0.5;ka=0.03;ki=0.06;`	
7	`c1=0.08;c2=-0.08;c3=0.1;c4=0.11;c5=0;` `c6=-0.15;`	Or, "c3=0.01 or c3=0.2"
8	`Apmax=0.2;Ipmax=0.5;`	
9	`Atemp=A(:,:,1); Itemp=I(:,:,1);`	
10	`Aright=Atemp;Aleft=Atemp;Aup=Atemp;` `Adown=Atemp;`	
11	`Iright=Itemp;Ileft=Itemp;Iup=Itemp;` `Idown=Itemp;`	
12	`for T=1:Tmax-1`	
13	`  Ap=c1*A(:,:,T)+c2*I(:,:,T)+c3;`	
14	`  Ip=c4*A(:,:,T)+c5*I(:,:,T)+c6;`	
15	`  Ap=((Ap<Apmax)&(Ap>0)).*Ap+(Ap>Apmax).` `*Apmax;`	Setting the upper and lower limits of Ap
16	`  Ip=((Ip<Ipmax)&(Ip>0)).*Ip+(Ip>Ipmax).` `*Ipmax;`	Setting the upper and lower limits of Ip
17	`  Atemp=A(:,:,T); Itemp=I(:,:,T);`	Calculating the diffusion of A and I
18	`  Aright(:,Xmax)=Atemp(:,Xmax);Aright(:,1:` `Xmax-1)=Atemp(:,2:Xmax);`	
19	`  Aleft(:,1)=Atemp(:,1);Aleft(:,2:Xmax)` `=Atemp(:,1:Xmax-1);`	
20	`  Aup(Xmax,:)=Atemp(Xmax,:);Aup(1:Xmax-1,:)` `=Atemp(2:Xmax,:);`	
21	`  Adown(1,:)=Atemp(1,:);Adown(2:Xmax,:)` `=Atemp(1:Xmax-1,:);`	
22	`  Iright(:,Xmax)=Itemp(:,Xmax);Iright(:,1:` `Xmax-1)=Itemp(:,2:Xmax);`	
23	`  Ileft(:,1)=Itemp(:,1);Ileft(:,2:Xmax)` `=Itemp(:,1:Xmax-1);`	
24	`  Iup(Xmax,:)=Itemp(Xmax,:);Iup(1:Xmax-1,:)` `=Itemp(2:Xmax,:);`	
25	`  Idown(1,:)=Itemp(1,:);Idown(2:Xmax,:)` `=Itemp(1:Xmax-1,:);`	
26	`  A(:,:,T+1)=dt*(da/dx2*(Aright+Aleft+Aup` `+Adown-4*Atemp)+Ap-ka*A(:,:,T))+A(:,:,T);`	
27	`  I(:,:,T+1)=dt*(di/dx2*(Iright+Ileft+Iup` `+Idown-4*Itemp)+Ip-ki*I(:,:,T))+I(:,:,T);`	
28	`end`	
29	`figure('Position',[0 300 1000 400]);`	

(continued)

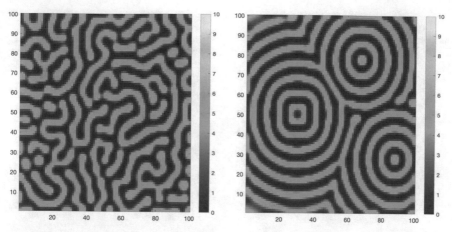

Fig. 4.31 Turing pattern using matrices (stripe)

30	`for T=1:100:Tmax`	
31	`  subplot(1,2,1);imagesc(A(:,:,T),[0 10]);` `axis xy;colorbar;`	
32	`  subplot(1,2,2);imagesc(I(:,:,T),[0 10]);` `axis xy;colorbar;`	
33	`  pause(0.0001);`	
34	`end`	

Lines 9–11 are just to prepare the matrices of Atemp, Aright, Aleft, Aup, Adown, etc. in advance and are not absolutely necessary. The result is shown in the left side of Fig. 4.31, and you can see that the problem of calculating the boundary region has been solved.

Now, change the initial conditions in line 3 to "A(46:55,26:35,1)=1;A (76:80,66:70,1)=1;A(26:30,86:90,1)=1;" (if Xmax=20, "A(9: 11,5:7,1)=1; A(15:18,13:14,1)=1;A(5:6,17:18,1)=1;"). This results in a pattern similar to the one shown in Fig. 4.31 (right), and we can see that it produces a striped pattern regardless of the initial conditions.

The Turing model can produce various patterns by changing a few parameters. Try changing the parameter in line 7 to "c3=0.01;." If we change the initial condition in line 3 to "A(:,:,1)=rand(Xmax,Xmax);," the result will be as shown in the left side of Fig. 4.32, and we can see that the parametertends to create spotlike patterns. Finally, if you set the parameter "c3=0.2;" in line 7, you will see a mesh-like pattern as shown in Fig. 4.33. You may want to try different values for the other parameters as well. If you have any questions about why such a variety of patterns are produced, you can read a paper on theoretical analysis [19].

The Turing model originally assumed two diffusible factors, but there was no evidence for the existence of such diffusible factors. On the other hand, recent studies by Dr. Shigeru Kondo and his colleagues at Osaka University have shown that the entities that make up the Turing pattern are two types of pigmented cells, and

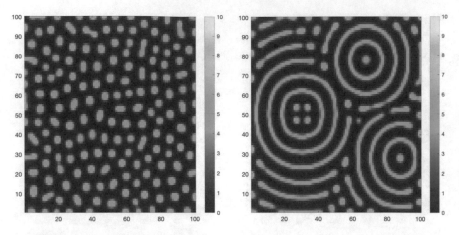

Fig. 4.32 Turing pattern using matrices (spot)

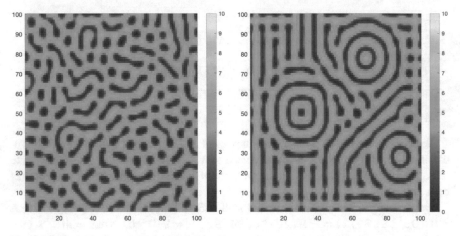

Fig. 4.33 Turing pattern using matrices (mesh)

that the signals used in their interaction are not diffusible, but that the information is transmitted directly when the projections of the cells come into contact with other cells [15, 16, 21]. So, was the Turing model wrong?

Theoretically, it is possible to achieve the same interaction as reaction–diffusion by interacting through the protrusions of cells. However, it is not necessary that the means of transmission be molecular diffusion, and as long as the conditions of **"activation at short distances and inhibition at long distances"** are met, the same pattern can be created even when information is transmitted by contact with cell protrusions [23, 24]. In this way, even if the physical entity is not clear in a phenomenological mathematical model, by combining experiments and developing theoretical research, we can clarify the mechanism by which patterns are generated at the molecular and cellular levels.

Exercise 4.7.4a Figure 4.25 shows photos of three kinds of fish: the snapper, the grouper, and the puffer fish. The patterns on each of these fish are similar to the patterns in Figs. 4.30, 4.31, and 4.32, especially for the grouper, which is almost identical to the spot pattern in Fig. 4.32 (left). For the other two, change the initial pattern of A without changing the parameters to make the pattern similar to that of the actual fish, as shown in Fig. 4.25.

Exercise 4.7.4b Instead of Eqs. (4.7.3a) and (4.7.3b), use the following equations to simulate the Turing model [20, 25]. $k_a = 1.2$, $c = 0.4$:

$$\frac{\partial A}{\partial t} = d_a \Delta A - k_a A + \frac{A^2}{I(1 + cA^2)} \tag{4.7.4a}$$

$$\frac{\partial I}{\partial t} = d_i \Delta I - k_i I + A^2 \tag{4.7.4b}$$

Note 8. Setting Parameters

There may be cases where a bug is caused by an improperly set parameter, which may not be a bug but is an important issue as well as debugging. In the case of the mathematical model of hematopoietic stem cells in Sect. 3.2.1, experimentally measured parameters could be used. But when it is difficult to measure parameters accurately, or when it is important to simulate parameters outside the range of measured parameters, it is necessary to calculate a wide range of parameters exhaustively.

You can rewrite the program for each parameter and execute it again, but you can also use a *for* statement to change the parameters themselves and display the results in a list, as in "script3_4D.m." In this case, we have prepared the parameter N as a vector, but you can also use the value of J changed by the *for* statement to get the value of N as "N=4*J." In this way, you can try as many different values of N as you like by changing the range of J.

If you want to try four different values of a parameter, you can display all the results in a graph as shown in Fig. 3.22. But if you have 100 different parameters, it will be difficult to display the results. In such a case, it is better to focus on the most important values in the results. If the total rate of infection is important, you can plot the value of N on the horizontal axis and the total rate of infection at time Tmax on the vertical axis (using *plot*). Then all the results can be represented in a single graph. Independently of N, it is also possible to vary the value of R_0 (discussed in Sect. 3.4.3) over a wide range and *plot* the total infection rate for various combinations of N and R_0 values (using *imagesc*). In this way, the effects of varying several parameters can be clearly

(continued)

seen. Such a diagram is called a **phase diagram**, and it is useful when discussing what parameter values are necessary for a phenomenon to occur.

When we calculated the production and degradation of proteins in "script3_1A.m," we assumed that the production rate of a protein is $p=1$ and the degradation rate coefficient is $k=0.1$. In the paper by Alber et al. [26], the production and degradation rates were estimated by attaching fluorescent tags to various proteins and applying the time variation of the fluorescence intensity to a mathematical model. These values vary depending on the type of protein and the time of the cell cycle, but roughly speaking, if the production rate is 50–100, the degradation rate is 0.05–0.1. The rate of synthesis is about 1000 times faster than that of degradation, as measured by the change in fluorescence intensity per hour. However, this value may be completely different for proteins that are actively degraded by the ubiquitin–proteasome system, as, for example, in [4], and it is necessary to change the parameters depending on the target protein of the mathematical model and the nature of the phenomenon being observed.

4.8 Simulation of Heartbeats

4.8.1 FitzHugh–Nagumo Equation

In the previous section, we have dealt with mathematical models focusing on pattern formation in the developmental process of organisms. However, there is no big difference between development and neural activity in terms of equations. It is the same in that it can be expressed by the reaction–diffusion equation [1].

Now, across the cell membrane of a neuron, the inside of the cell has a lower potential than the outside (**resting membrane potential**; Fig. 4.34). When a **sodium channel** opens and **sodium ions (Na+)** flow into the cell, the membrane potential rises. When the membrane potential rises above a certain **threshold**, the influx of sodium ions is accelerated, and the rapid increase in the membrane potential generates an **action potential**. In turn, **potassium channels** open and **potassium ions (K+)** flow out, causing the membrane potential to decrease and return to the resting membrane potential.

The mechanism of action potential generation was clarified by Hodgkin and Huxley, and they further developed a mathematical model of action potential generation by sodium and potassium channels (**Hodgkin–Huxley model**)

Fig. 4.34 Changes in membrane potential

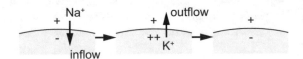

[27]. However, this mathematical model is complex, consisting of four variables and including many parameters.

On the other hand, the **FitzHugh–Nagumo equation** is well known as a simplified model that shows the same behavior as the Hodgkin–Huxley model (Eqs. (4.8.1a) and (4.8.1b)) [28, 29]. This is a mathematical model consisting of two variables, but it is useful in considering the mechanism of action potential generation [1]. Here, we consider the variable V, which represents the membrane potential. During the **refractory period** immediately after the action potential is generated, sodium channels become inactive, and the membrane potential does not rise. The variable W represents the refractoriness of the membrane potential:

$$\frac{dV}{dt} = c\left(-\frac{V^3}{3} + V - W + I \right) \tag{4.8.1a}$$

$$\frac{dW}{dt} = (V - bW + a)\frac{1}{c} \tag{4.8.1b}$$

Here, we use a set of general parameters, $a=0.7$, $b=0.8$, and $c=10$. I is the input current, which represents the current artificially injected from an electrode or excitatory input from another neuron. Let us assume that the input current is always $I=1$. Write a program that plots the temporal variation of V and W in a single cell with dt=0.02 and Tmax=10000, without considering the spatial extent. It will be as follows:

script4_8A.m

```
1    Tmax=10000;dt=0.02;
2    V=zeros(1,Tmax);W=zeros(1,Tmax);
3    a=0.7;b=0.8;c=10;I=1;
4    for T=1:Tmax-1
5      V(T+1)=V(T)+dt*(c*(-(V(T).^3)/3+V(T)-W(T)+I)));
6      W(T+1)=W(T)+dt*(V(T)-b*W(T)+a)/c;
7    end
8    plot(1:Tmax,V,'r',1:Tmax,W,'k');
```

When calculating the cube of V (V^3) in line 5, we need to put a period (.) before "^," just as in the Hadamard product, because we are multiplying the value of each element.

V is plotted as a red line, and W as a black line. We can see that W rises as V rises, followed by a sharp drop in V. Then, as V drops, W drops, and V rises again (Fig. 4.35). In this way, we can see that the generation of action potentials is repeated.

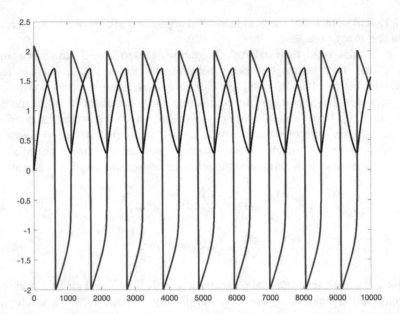

Fig. 4.35 Calculation result of FitzHugh–Nagumo equation

4.8.2 Adding Diffusion to FitzHugh–Nagumo Equation

This model can be extended to simulate a beating heart [30]. **Cardiac muscle** exhibits membrane potential changes similar to those of nerve cells and can be described by the FitzHugh–Nagumo equation. On the other hand, since the cells of the myocardium are electrically connected to each other by a structure called **gap junctions**, current flows in proportion to the difference in membrane potential between neighboring cells, and consequently, the membrane potential V has the effect of diffusing and spreading between the cells. Similarly, the membrane refractoriness W also has a spatial spreading effect, so we can redefine them as $V(t, x, y)$ and $W(t, x, y)$. However, we will omit the (t, x, y) part in the following equations and refer to them as "$V(Y,X,T)$" and "$W(Y,X,T)$" in the MATLAB script.

Figure 4.36 shows a schematic of electrical stimulation propagation through the heart and myocardium. First, the **sinus node** receives a constant current input I from the **autonomic nerve**. The electric potential V propagates through the **atria**, causing the atrial myocardium cells to contract one after another. When this stimulus travels to the **atrioventricular node**, the electrical stimulus now travels to the **ventricular myocardium**, causing the atria and ventricles to contract with a time delay.

To simplify our discussion, we will focus on the sinus node to the atria and omit the propagation of the stimulus from the atrioventricular node to the ventricles. We will also consider the two-dimensional expansion of the myocardium on the surface of the atria, looking at the sinus node as the center. Adding the diffusion of

Fig. 4.36 Propagation of
the stimulus from sinus node
to atria

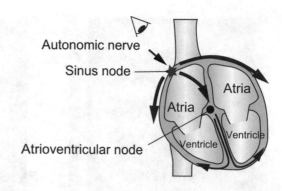

V (diffusion coefficient $d = 1$) to Eq. (4.8.1a), we get the following equation. The
parameters are a=0.7, b=0.8, c=10.

$$\frac{\partial V}{\partial t} = c\left(-\frac{V^3}{3} + V - W + I \right) + d\Delta V \qquad (4.8.2a)$$

$$\frac{\partial W}{\partial t} = (V - bW + a)\frac{1}{c} \qquad (4.8.2b)$$

Assume that dt=0.02, dx=1, Tmax=10000, Xmax=100, and the current from
the sinus node is constant, I=1, in the central region of X=50:51, Y=50:51, and
"I=zeros(Xmax,Xmax);I(50:51,50:51)=1;." Let the sinus node be the
center of a two-dimensional plane with 100-by-100 myocardium, and calculate the
potential change of the myocardium. If we assume that the myocardium contracts
when the electric potential increases, we can simulate the contraction pattern of the
myocardium with a program like "script4_8B.m" below:

script4_8B.m		
1	`Tmax=10000;Xmax=100;`	Or, "Tmax=5000;"
2	`dt=0.02;dx=1;dx2=dx*dx;`	
3	`V=zeros(Xmax,Xmax,Tmax);W=zeros(Xmax,Xmax,Tmax);`	
4	`I=zeros(Xmax,Xmax);I(50:51,50:51)=1;`	Setting the spatial distribution pattern of I
5	`a=0.7; b=0.8; c=10;d=1;`	
6	`for T=1:Tmax-1`	
7	`V(:,:,T+1)=V(:,:,T)+dt*(c*(-(V(:,:,T).^3)/3+V(:,:,T)`	
	`-W(:,:,T)+I) + 4*d*del2(V(:,:,T)) );`	
8	`W(:,:,T+1)=W(:,:,T)+dt*(V(:,:,T)-b*W(:,:,T)+a)/c;`	

(continued)

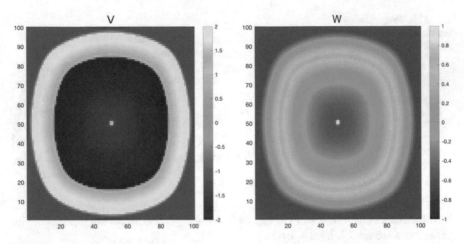

Fig. 4.37 Simulation of heartbeats

9	end	
10	figure('Position',[0 300 1000 400]);	
11	for T=1:100:Tmax	
12	subplot(1,2,1);imagesc(V(:,:,T),[-2 2]);axis xy;colorbar;	
13	subplot(1,2,2);imagesc(W(:,:,T),[-1 1]);axis xy;colorbar;	
14	pause(0.0001);	
15	end	

Unlike "script4_8A.m," V and W are three-dimensional matrices, and in lines 7–8, by fixing the time to T or T+1, we consider them as two-dimensional matrices. Similarly, in lines 12–13, *imagesc* is used to display patterns of 2D matrices such as "V(:,:,T)" and "W(:,:,T)" (see Fig. 2.31). In "script4_8A.m," V and W are vectors (1D matrices), so it is possible to replace V(1,T) with V(T) or V(1,:) with V. However, in the case of 3D matrices, such abbreviations will not work properly.

The pattern shown in Fig. 4.37 propagates from the center to the outside, which means that the myocardium sequentially contracts in order from the sinus node area. Please try to replace *del2* with a faster version using matrices by yourself.

4.8.3 Numerical Simulation of Arrhythmia

In the mathematical model described in the previous section, the propagation of electrical excitation in cardiomyocytes through gap junctions was introduced in the form of diffusion of potential *V*. When the propagation of excitation is impaired due

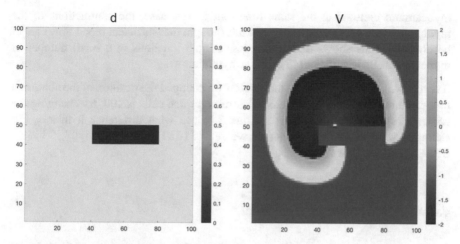

Fig. 4.38 Simulation of arrhythmia

to abnormalities in these gap junctions, it is known that the heart beats incompletely, resulting in **arrhythmia** [31, 32]. Although it is obvious that arrhythmia would occur if all cells were dysfunctional, we will now consider the case where the gap junctions of a small number of cells are abnormal and electrical coupling with neighboring cells is impaired.

In the previous section, we assumed that the diffusion coefficient $d = 1$, but when the electrical coupling is impaired, the membrane potential is not transmitted to the surrounding cells, so we assume $d = 0$. Here, we define the diffusion coefficient d as a 100-by-100 two-dimensional matrix and consider the case where 100 out of 10,000 cardiomyocytes (1% of the total) are abnormal.

For example, if "d=ones(100,100);d(41:50,41:80)=0;," then d will have the pattern shown in the left side of Fig. 4.38. We can replace "d=1;" in line 5 of "script4_8B.m" with this, but there is one more correction that needs to be made. In line 8, "+4*d*del2(V(:,:,T))," the scalar d is now a two-dimensional matrix, so the product of d and del2(V) is not the regular product, but the Hadamard product, which needs to be ".*." Then we have the following:

script4_8C.m: modified from "script4_8B.m"

5	a=0.7; b=0.8; c=10; d=ones(100,100);d (41:50,41:80)=0;	Setting the spatial distribution of d
7	V(:,:,T+1)=V(:,:,T)+dt*(c*(-(V(:,:, T).^3)/3+V(:,:,T)	
	-W(:,:,T)+I) + 4*d.*del2(V(:,:,T)));	Hadamard product is used since d is a 2D matrix

This modification results in an irregular propagation pattern as shown in Fig. 4.38 (only the value of V is shown in the figure). Normally, concentric circles of the myocardium contract at the same time, but in this case, the contractions of the myocardium are not synchronized, and it is thought that sufficient blood flow cannot be obtained. This shows that even a loss of gap junctions in a small number of myocardial cells is sufficient to cause arrhythmia.

Exercise 4.8.3 What would be the result if we changed the position of the abnormal myocardium? Assuming that the number of abnormal cells is 100, try changing the contents of the matrix d defined in line 5 to see what difference it makes. For example, what results do we get, if we set "d(46:55,46:55)=0", "d(51: 55,61:80)=0", or "d(41:50,61:80)=0"?

4.8.4   Mechanism of Heartbeat

Up to now, we have been using computers to perform numerical calculations based on mathematical models to study the behavior of the systems we are focusing on. However, it is also important to analyze the general properties of the equations that constitute the mathematical model itself. Here, I would like to discuss how action potentials are generated in the FitzHugh–Nagumo equation. For details, please refer to more specialized books, for example, [3]. This section may be a little difficult, so you may skip it.

When discussing the properties of a mathematical model, it is important to focus on certain characteristic states. First, we will simplify Eqs. (4.8.2a) and (4.8.2b) and consider a cell with no current input ($I = 0$). Also, we consider only one cell, ignoring its spatial extent ($d\Delta V$ is ignored). Then, consider the steady state of V and W, that is, a constant state that does not change with time ($\frac{dV}{dt} = \frac{dW}{dt} = 0$). Then, you obtain the following:

$$\frac{dV}{dt} = -\frac{V^3}{3} + V - W = 0 \qquad (4.8.3)$$

$$\frac{dW}{dt} = V - bW + a = 0 \qquad (4.8.4)$$

If we transform them further, we obtain

$$W = -\frac{V}{3}(V + \sqrt{3})(V - \sqrt{3}) \qquad (4.8.5)$$

$$W = \frac{1}{b}(V + a) \qquad (4.8.6)$$

Equations (4.8.5) and (4.8.6) show a set of points where V and W are time-invariant, respectively, and such a set is called a **nullcline**. At the **equilibrium point**, which is the intersection of the two null lines (the red point in Fig. 4.39a), both V and W are time-invariant. Plotting them on a two-dimensional plane, we can divide the

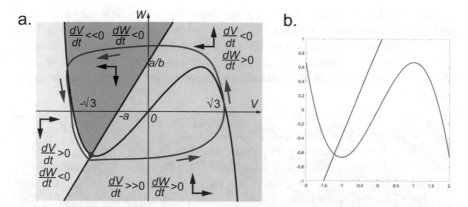

Fig. 4.39 A vector field of FitzHugh–Nagumo equation. (**a**) A phase plane with a vector field. (**b**) A phase plane drawn on MATLAB

plane into four regions: yellow, pink, green, and blue, as shown in Fig. 4.39a. Note that across the null line, the positive and negative values of $\frac{dV}{dt}$ and $\frac{dW}{dt}$ are reversed. In the pink and green regions, $\frac{dW}{dt}$ becomes positive in the region below the line in Eq. (4.8.6), so W will continue to increase with time in this region. Conversely, in the yellow and blue regions, W will continue to decrease.

On the other hand, in the yellow/pink region, which is below the curve in Eq. 4.8.5, $\frac{dV}{dt}$ will be positive and V will continue to increase. Conversely, in the green and blue regions, V continues to decrease. Also, since Eq. (4.8.3) has a term for the cube of V, we can expect the value to change faster along the V-axis than along the W-axis at a distance from the curve in Eq. (4.8.5).

The diagram that represents such a property of a mathematical model is called a **phase plane**, in which we can draw a **vector field** that illustrates how the values of each variable change in each region using vectors based on the equilibrium point and the nullcline. This allows us to explain the change in values as if we were rotating leftward on the magenta line shown in Fig. 4.39a. This type of analysis is called **phase-planar analysis**, and it is useful for understanding how the phenomenon we are focusing on arises.

Exercise 4.8.4 Draw a diagram shown in Fig. 4.39b modifying "script4_8B.m" ("script4_8D.m").

script4_8D.m : modified from "script4_8B.m"

5	`a=0.7;b=0.8;c=10;d=1;X=-2:0.01:2;` `Y1=-(X.^3)/3+X;Y2=(X+a)/b;`	Preparing vectors X, Y1, Y2 to plot the nullclines
13	`subplot(1,2,2);plot(X,Y1,X,Y2,V(71,51,` `T),W(71,51,T),'*');`	
	`ylim([-1 1]);`	Plot the values of V, W with asterisks

Here, Y1 represents Eq. (4.8.5), Y2 represents Eq. (4.8.6), and the values of V and W at the point where Y=71, X=51 are plotted with asterisks. If line 11 is "for T=1:100:Tmax," T increases by 100 and it is too fast to see the animation of the asterisk. It would be better to change it to something like "for T=1000:10: Tmax." You can see that the behavior is similar to the magenta line shown in Fig. 4.39a.

Note 9. Values of *dt* and *dx*

In "script3_5A.m," we have shown that the results can vary greatly depending on the value of *dt*, but in this case, we thought that $dt = 0.002$ was more reliable than $dt = 0.01$ because it was closer to the results obtained using *ode45*. However, *ode45* does not calculate an exact solution but a numerical solution. And it is a function for ordinary differentiation and is not applicable to partial differential equations. In the Euler method, as the value of *dt* decreases, the result becomes closer to the exact solution, so you can decrease the value of *dt* little by little and judge that *dt* is small enough if the result is almost the same.

The Euler method described in Sect. 3.1.1 has the problem that the solution is not stable depending on the value of *dt*. It is known that more complicated methods can stabilize the solution even if the value of *dt* is somewhat significant, so please refer to other references for details [2, 3].

Biological or biochemical considerations are useful not only for estimating parameters but also for examining the values of *dt* and *dx*. "script3_1A.m" calculates the time variation of the concentration of protein E as shown in Fig. 3.1, where $dt = 0.01$, production rate $p = 1$, and degradation rate $k = 0.1$. E approaches 10. This value of 10 is the maximum concentration of E in the cell. In the equation in Fig. 3.1, $p \times dt$ is the amount of E produced per micro time. If $dt = 10$, then $p \times dt = 10$, which means that E reaches its maximum concentration in just one micro time. This is too simplistic. If $dt = 1$, then $p \times dt = 1$, and the maximum concentration will be reached in just 10 micro times, so we can assume that it would be better to keep $dt = 0.1$ or less at least.

For the diffusion equation, we need to consider not only *dt* but also the value of the small distance *dx*. The conditions for stable diffusion calculations $\frac{d\,dt}{dx^2} < \frac{1}{4}$ were discussed in Sect. 4.5.1, but *dt*, *dx*, and the diffusion coefficient *d* are interrelated. Let's consider the condition of *dx* when $dt = 0.01$. The diffusion coefficient *d* means the area where molecules diffuse per unit time. It has been reported that the diffusion coefficient of fluorescently tagged Fgf8 in zebrafish embryos is 4–53 μm^2/sec [33]. Assuming that the diffusion coefficients of diffusible materials are generally the same, the appropriate range of dx^2 should be approximately $dx^2 > 0.2$–2.0 μm^2.

Fig. 4.40 Signal
transduction of EGF and
Notch

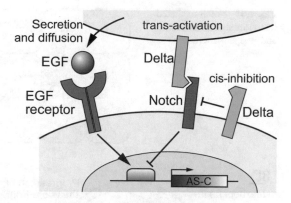

4.9   Mathematical Model of the Wave of Differentiation

4.9.1   The Proneural Wave

I believe that you can construct various mathematical models and obtain results by calculating them by applying the previous examples. In my laboratory, we focus on a phenomenon called the "Proneural wave," which is a wave of differentiation observed during the formation of the fly brain, and constructed a mathematical model of the proneural wave in order to elucidate its propagation mechanism [34–39].[3] The important point is that we have not only reproduced the phenomenon using the mathematical model but also found an interesting new phenomenon inspired by the mathematical model and shown that it can be observed in actual living organisms. It was not only a simulation but also a driving force in revealing new mechanisms of life and creating substantial progress in life science. In this section, I would like to explain the questions that led us to the mathematical modeling of proneural waves and how we finally tested the hypotheses derived from the model through experiments.

We have considered cell-to-cell signaling through secretory ligands, but there are cases in which the ligands are not secretory but membrane-bound (Fig. 4.1). For example, in **Notch** signaling, the Notch receptor is activated by binding of a membrane-bound ligand, **Delta**, to regulate the expression of target genes (Fig. 4.40). Notch and Delta are universally important proteins shared by not only flies but also mammals, including humans [4].

Secretory ligands diffuse between cells, whereas membrane-bound ligands remain at the plasma membrane of the expressing cell. Therefore, Delta expressed in differentiated cells can only activate Notch in neighboring cells, as shown in

---

[3]This phenomenon was originally discovered in the laboratory of Dr. Tetsuya Tabata at the Institute for Quantitative Life Sciences, the University of Tokyo, and its molecular mechanism has been elucidated. The mathematical model is the result of collaborative research with Dr. Masaharu Nagayama of Hokkaido University and Dr. Takashi Miura of Kyushu University.

Fig. 4.41 Notch-mediated lateral inhibition and salt-and-pepper pattern

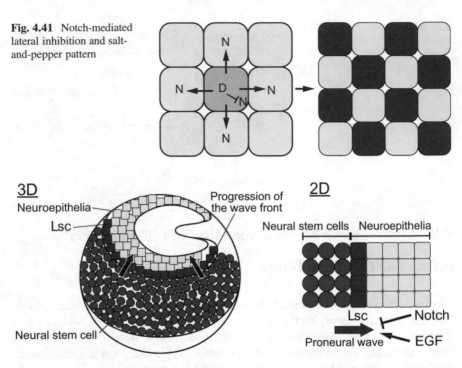

Fig. 4.42 Propagation of the proneural wave

Figs. 4.40 and 4.41. Notch activation suppresses cell differentiation in Notch-activated cells, and Delta expression is suppressed in undifferentiated cells, resulting in a salt-and-pepper pattern in which only a subset of cells differentiates (Fig. 4.41). Such a function of Notch is called **lateral inhibition**. Additionally, when Delta is expressed in the same cells as Notch, it inhibits Notch activity, and this effect is called *cis*-**inhibition** (Figs. 4.40 and 4.41).

Secretory ligands are responsible for long-range signaling by diffusion, while membrane-bound ligands are responsible for short-range signaling by cell–cell adhesion. What would happen if the two coexisted and worked in harmony?

The proneural wave, which is observed during the formation of the fly brain, is an ideal system to consider such a question. As shown in Figs. 4.41 and 4.42, the developing fly brain initially contains only undifferentiated neuroepithelial cells, which in turn differentiate row by row into neural stem cells [36, 37].

The progression of the proneural wave is promoted by the secreted ligand of EGF and inhibited by Delta/Notch (Figs. 4.43, 4.44, and 4.45) [34, 35]. EGF stands for epidermal growth factor, which is also a universally important protein common to not only flies but also various animals [4]. In flies, it is easy to mutate a certain gene in a part of the body by a method called mosaic analysis. The results of mosaic analyses demonstrated that the wave of differentiation disappears in the region of EGF mutants (Fig. 4.44) and accelerates in the region of Notch mutants (Fig. 4.45).

Fig. 4.43 The proneural wave

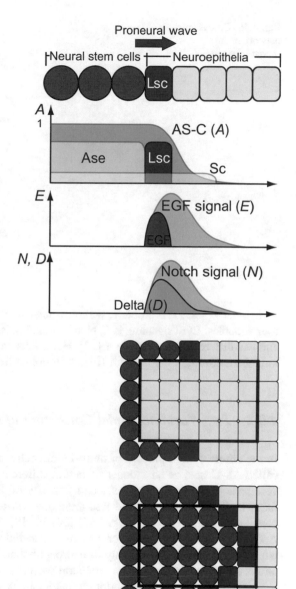

Fig. 4.44 EGF mutant cells

Fig. 4.45 Notch mutant cells

Notch-induced lateral inhibition is a common phenomenon in diverse biological phenomena in all animals, including the proneural wave. Notch-induced lateral inhibition should result in the salt-and-pepper pattern shown in Fig. 4.41. However, such a pattern is not observed in the proneural wave (Fig. 4.42). Why does Notch not form a salt-and-pepper pattern but instead suppress the progression of the wave, a seemingly completely different function?

Fig. 4.46 Gene regulatory
network of the
proneural wave

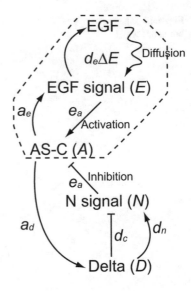

In order to solve this problem, we constructed a mathematical model consisting of four variables: EGF signaling is E, Notch signaling is N, Delta expression level is D, and cell differentiation level is A [37]. Here, let's start with a simple mathematical model considering only E and A (Fig. 4.46 dotted line).

4.9.2   Two-Variable Model Consisting of EGF and AS-C

In the differentiation process of **neural stem cells**, a group of transcription factors called **AS-C** acts as an inducer of neural differentiation. AS-C is a homolog of vertebrate Neurogenin and Ascl and consists of four homologous transcription factors, of which Lsc, Sc, and Ase are expressed to differentiate **neuroepithelial cells** into neural stem cells (Fig. 4.43) [34, 36, 38]. The expression of Lsc, Sc, and Ase results in the differentiation of neuroepithelial cells into neural stem cells. In particular, Lsc is expressed only in neuroepithelial cells just before differentiation and is useful as a marker of the proneural wave.

Although these AS-C transcription factors are thought to induce cell differentiation, the process of cell differentiation is not simply determined by the expression levels of differentiation-inducing factors but also by **epigenetic** mechanisms such as the global state of **chromosomes** [4]. Since it is very difficult to make a rigorous mathematical model of cell differentiation, it is sufficient to reproduce the phenomenon of differentiation phenomenologically.

It is generally believed that cell differentiation is an irreversible process and that once a cell has differentiated, it will never return to its **undifferentiated** state [4]. In fact, neural stem cells that differentiate from neuroepithelial cells immediately begin to divide asymmetrically, producing a large number of neurons toward the inner side

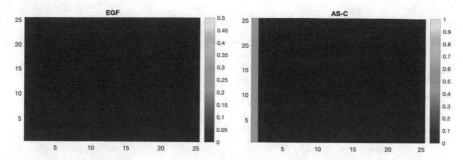

Fig. 4.47 Initial distributions of E and A

of the brain. Therefore, we assume that the variable A is not a physical quantity of AS-C but an abstract quantity of cell differentiation: $A = 0$ means undifferentiated neuroepithelium, $A = 1$ means differentiated neural stem cells, and once $A = 1$, the value of A does not change further.

Since EGF signaling promotes cell differentiation, how can we ensure that the time course of A depends on E and that A no longer changes when $A = 1$? The simplest equation is as follows:

$$\frac{\partial A}{\partial t} = e_a(1-A)E \qquad (4.9.1)$$

If A is greater than or equal to 0 and less than 1, the time variation of A will be proportional to E (coefficient e_a), but if $A = 1$, A will not change with time because $1 - A = 0$. Of course, there are many other possible functions whose right-hand side becomes zero when $A = 1$, but it is best to use the simplest equation you can think of and improve the mathematical model if problems arise.

How can we describe the change in E over time, where E is diffusing, degrading, and positively regulated by A (Fig. 4.46)? However, since EGF is produced by undifferentiated neuroepithelial cells and not by differentiated neural stem cells (Fig. 4.43), and since E does not increase when $A = 1$, we can use the same method as in Eq. (4.9.1) to obtain

$$\frac{\partial E}{\partial t} = d_e\Delta E - k_eE + a_eA(1-A) \qquad (4.9.2)$$

d_e is the diffusion coefficient of E, k_e is the degradation rate coefficient, and a_e is the production rate coefficient of E by A.

Now, let's set $T_{max} = 200$, $X_{max} = 25$, $dt = 0.1$, $dx = 1$, $d_e = 1$, $k_e = 1$, $a_e = 1$, $e_a = 10$ and create a program to calculate the mathematical model in Eqs. (4.9.1) and (4.9.2). We assume that the reaction rates of all chemical reactions are equivalent and set most of the parameters to 1, but e_a is set to a larger value of 10 because it affects the reaction rate of the entire system.

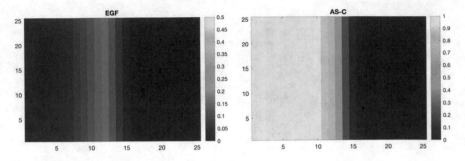

Fig. 4.48 Results of two-variable model

When T=1, "E(:,:,1)=0" and "A(:,1,1)=0.5" (Fig. 4.47). This corresponds to the experimental result that the leftmost cell expresses Lsc and differentiates a little before the proneural wave actually progresses [37]. *del2* does not work well for the calculation of the mathematical model of the proneural wave, so we use a fast matrix-based calculation for diffusion from the beginning as shown in "script4_9A.m." You can see that the differentiation wave moves to the right with time (Fig. 4.48).

	script4_9A.m	
1	`Tmax=200;Xmax=25;`	
2	`dt=0.1;dx=1;dx2=dx*dx;`	Setting the values of dt, dx, dx^2
3	`de=1;ke=1;ae=1;ea=10;`	
4	`E=zeros(Xmax,Xmax,Tmax);`	
5	`A=zeros(Xmax,Xmax,Tmax);A(:,1,1)=0.5;`	Setting the initial pattern of differentiation, A
6	`Etemp=zeros(Xmax,Xmax);Eright=Etemp;` `Eleft=Etemp;Eup=Etemp;Edown=Etemp;`	
7	`for T=1:Tmax-1`	
8	`Etemp=E(:,:,T);`	Calculating the reaction–diffusion of E
9	`Eright(:,Xmax)=Etemp(:,Xmax);` `Eright(:,1:Xmax-1)=Etemp(:,2:Xmax);`	
10	`Eleft(:,1)=Etemp(:,1);Eleft(:,2:Xmax)` `=Etemp(:,1:Xmax-1);`	
11	`Eup(Xmax,:)=Etemp(Xmax,:);Eup(1:Xmax-` `1,:)=Etemp(2:Xmax,:);`	
12	`Edown(1,:)=Etemp(1,:);Edown(2:Xmax,:)` `=Etemp(1:Xmax-1,:);`	

(continued)

13	`E(:,:,T+1)=dt*(de*(Eright+Eleft+Eup` `+Edown-4*Etemp)/dx2`	
	`-ke*E(:,:,T)+ae*A(:,:,T).*(1-A(:,:,` `T)))+E(:,:,T);`	
14	`A(:,:,T+1)=dt*(ea*(1-A(:,:,T)).*E(:,:,` `T))+A(:,:,T);`	Calculating the time difference of A
15	`end`	
16	`figure('Position',[0 400 1000 250]);`	
17	`for T=1:10:100`	Adjust the range of T for animation
18	`subplot(1,2,1);imagesc(E(:,:,T),` `[0 0.5]);axis xy;title('EGF');colorbar;`	
19	`subplot(1,2,2);imagesc(A(:,:,T),[0 1]);` `axis xy;title('AS-C');colorbar;`	
20	`pause(0.001);`	
21	`end`	

The reason of "`for T=1:10:100`" in line 17 is to display the image every 10 frames and stop the animation when the proneural wave passes the center of the region (T=100). You can change it as you like. For example, if you set "`for T=1:10:Tmax`," it will play until the end.

In this example, we set "Xmax=25" and consider a two-dimensional plane with 25-by-25 cells. The micro length dx represents the length of the side of each cell, and the diffusion $d_e \Delta E = d_e \left(\frac{\partial^2}{\partial x^2} + \frac{\partial^2}{\partial y^2} \right) E$ means that EGF diffuses in units of dx, which is the size of the cell. This is inaccurate because EGF actually diffuses out of the cell, not moves into each cell unit. EGF should diffuse on a finer spatial scale, but as we can see from Eq. (4.9.1), if a cell receives even a small amount of EGF, A will increase and the cell itself will produce EGF (Eq. (4.9.2)). Even if EGF diffuses on a finer spatial scale, the results will not be very different. In fact, we have confirmed that the same results are obtained when EGF diffuses on a finer spatial scale [37, 39].

The actual paper uses a colormap that is blue for small values, changes to light blue-green-yellow as the value increases, and becomes red at the maximum value [37]. "`colormap(jet);`" can be added after the *figure* command in line 16 to get the same colormap as in the paper.

4.9.3   A Four-Variable Model with Notch and Delta

Now, since our original goal was to see what happens when Notch-mediated lateral inhibition is added to EGF-induced reaction–diffusion, let's add Notch (N) and Delta (D) to the above model. D is degraded and positively regulated by A (Fig. 4.46). However, D, like E, is produced by undifferentiated neuroepithelial cells and not by differentiated neural stem cells (Fig. 4.43). Since D does not increase when $A = 1$, as in Eq. (4.9.2), we have

$$\frac{\partial D}{\partial t} = -k_d D + a_d A(1 - A) \tag{4.9.3}$$

k_d is the degradation rate coefficient of D, and a_d is the production rate coefficient of D by A.

N represents the activity of the Notch signal, which is spontaneously degraded and activated by D expressed in neighboring cells and conversely inhibited by D from the same cell (Figs. 4.40 and 4.41). The equation for this is as follows:

$$\frac{\partial N}{\partial t} = -k_n N + d_n \sum_{l,m} D_{l,m} - d_c D_{i,j} \tag{4.9.4}$$

$\sum_{l,m} D_{l,m}$ is the sum of the amounts of D expressed in neighboring cells (Fig. 4.41). k_n is the degradation rate coefficient of N, d_n is the coefficient of N activation by D in neighboring cells (***trans*-activation**), and d_c is the coefficient of N inhibition by D in the same cell (***cis*-inhibition**). Let's write the code based on Eqs. (4.9.1)–(4.9.4) and the parameters we just discussed, and also setting $k_n = 1$, $d_n = 0.25$, $d_c = 0.25$, $k_d = 1$, $a_d = 1$. Again, all chemical reactions are assumed to be equivalent, but the *trans*-activation (d_n) of N by D is usually the sum of the inputs from four neighboring cells, so we set $1/4 = 0.25$. In addition, *cis*-inhibition is assumed to be equivalent to *trans*-activation, and $d_c = 0.25$ is used.

Note that in Eq. (4.9.4), the value of *cis*-inhibition is $-d_c D_{i,j}$, which may cause the value of N to be negative in some cases. This is fine for the parameters used in this book, but you can avoid the situation where N becomes negative by setting it to $-d_c D_{i,j} N$ [39].

Equation (4.9.2) remains unchanged, but Eq. (4.9.1) needs to be modified according to Fig. 4.46. In other words, considering that A is positively regulated by E and negatively regulated by N, and that cell differentiation is irreversible, we have the following:

$$\frac{\partial A}{\partial t} = e_a(1 - A) \max\{E - N, 0\} \tag{4.9.5}$$

where $\max\{E - N, 0\}$ means use the larger of "$E - N$" and "0," so that differentiated cells do not revert to the undifferentiated state even if $E - N$ becomes negative.

Now, we add Eqs. (4.9.3), (4.9.4), and (4.9.5) to "script4_9A.m" to complete "script4_9B.m." Equation (4.9.3) is almost the same as Eq. (4.9.2) and is written as "D(:,:,T+1)=dt*(-kd*D(:,:,T)+ad*A(:,:,T).*(1-A(:,::,T))) +D(:,:,T);" in the MATLAB code.

About Eq. (4.9.4), for cells outside the boundary region of "X=2:Xmax-1", "Y=2:Xmax-1," we can simply add the values of D in the four neighboring cells.

In lines 7–31 of "script4_3B.m," we modified the range of the *for* statement to handle the calculation of the boundary region. This same method can be used to solve the problem.

What should we do about max$\{E - N, 0\}$ in Eq. (4.9.5)? First, prepare a two-dimensional matrix named *En*, and set $En = E - N$. If any element of *En* is greater than 0, set the original value; otherwise, set it to 0. In other words, you can use the same idea as in line 9 of "script4_7C.m." Here, "(En>0)" is a matrix whose elements are 1 if the value is greater than 0, and 0 otherwise. So, we can just set "En . * (En>0)." Note that the product here is the Hadamard product. If you pay attention to the above points, the code will look like:

script4_9B.m

1	`Tmax=200;Xmax=25;`	
2	`dt=0.1;dx=1;dx2=dx*dx;`	
3	`de=1;ke=1;ae=1;ea=10;kn=1;dn=0.25;` `dc=0.25;kd=1;ad=1;`	
4	`E=zeros(Xmax,Xmax,Tmax);En=zeros(Xmax,` `Xmax);`	Preparing a 2D matrix En
5	`N=zeros(Xmax,Xmax,Tmax);D=zeros(Xmax,` `Xmax,Tmax);`	
6	`A=zeros(Xmax,Xmax,Tmax);A(:,1,1)=0.5;`	
7	`Etemp=zeros(Xmax,Xmax);Eright=Etemp;` `Eleft=Etemp;Eup=Etemp;Edown=Etemp;`	
8	`for T=1:Tmax-1`	
9	`  Etemp=E(:,:,T);`	Calculating the reaction–diffusion of E in lines 9–14
10	`  Eright(:,Xmax)=Etemp(:,Xmax);Eright` `(:,1:Xmax-1)=Etemp(:,2:Xmax);`	
11	`  Eleft(:,1)=Etemp(:,1);Eleft(:,2:Xmax)` `=Etemp(:,1:Xmax-1);`	
12	`  Eup(Xmax,:)=Etemp(Xmax,:);Eup(1:Xmax-` `1,:)=Etemp(2:Xmax,:);`	
13	`  Edown(1,:)=Etemp(1,:);Edown(2:Xmax,:)` `=Etemp(1:Xmax-1,:);`	

14	`  E(:,:,T+1)=dt*(de*(Eright+Eleft+Eup`
	`+Edown-4*Etemp)/dx2`
	`    -ke*E(:,:,T)+ae*A(:,:,T).*(1-A(:,:,`
	`T)))+E(:,:,T);`	

15	`  for X=2:Xmax-1`	Calculating the lateral inhibition of N
16	`    for Y=2:Xmax-1`	The region of X=2:24, Y=2:24 in line 15–31

(continued)

17	`N(Y,X,T+1)=N(Y,X,T) + dt*( dn*( D(Y,X +1,T)+D(Y,X-1,T)`	
	`+D(Y+1,X,T)+D(Y-1,X,T)) -dc*D(Y,X,T) -kn*N(Y,X,T) );`	
18	`end`	
19	`end`	
20	`for X=2: Xmax-1`	The top and bottom regions in line 20–23
21	`N(1,X,T+1)= N(1,X,T) + dt*( dn*( D(1,X +1,T)+D(1,X-1,T)`	
	`+D(2,X,T)) -dc*D(1,X,T) -kn*N(1,X, T));`	The region of Y=1
22	`N(Xmax,X,T+1)=N(Xmax,X,T) + dt*( dn*(D (Xmax,X+1,T)`	
	`+D(Xmax,X-1,T)+D(Xmax-1,X,T))-dc*D (Xmax,X,T) -kn*N(Xmax,X,T) );`	The region of Y=25
23	`end`	
24	`for Y=2: Xmax-1`	The left and right regions in line 24–27
25	`N(Y,1,T+1)= N(Y,1,T) + dt*( dn*( D(Y +1,1,T)+D(Y-1,1,T)`	
	`+D(Y,2,T)) -dc*D(Y,1,T) -kn*N(Y,1,T) );`	The region of X=1
26	`N(Y,Xmax,T+1)= N(Y,Xmax,T) + dt*( dn*( D (Y+1,Xmax,T)`	
	`+D(Y-1,Xmax,T)+D(Y,Xmax-1,T))-dc*D (Y,Xmax,T) -kn*N(Y,Xmax,T) );`	The region of X=25
27	`end`	
28	`N(1,1,T+1)=N(1,1,T) + dt*( dn*( D(1,2,T) +D(2,1,T))`	
	`-dc*D(1,1,T)-kn*N(1,1,T));`	The region of X=1, Y=1
29	`N(Xmax,1,T+1)= N(Xmax,1,T) + dt*(dn*(D (Xmax,2,T)+D(Xmax-1,1,T))`	
	`-dc*D(Xmax,1,T)-kn*N(Xmax,1,T));`	The region of X=25, Y=1
30	`N(1,Xmax,T+1)= N(1,Xmax,T) + dt*(dn*(D (2,Xmax,T)+D(1,Xmax-1,T))`	
	`-dc*D(1,Xmax,T)-kn*N(1,Xmax,T));`	The region of X=1, Y=25
31	`N(Xmax,Xmax,T+1)= N(Xmax,Xmax,T) +dt* (dn*(D(Xmax,Xmax-1,T)`	
	`+D(Xmax-1,Xmax,T))-dc*D(Xmax,Xmax,T) - kn*N(Xmax,Xmax,T));`	The region of X=25, Y=25
32	`D(:,:,T+1)=dt*(-kd*D(:,:,T)+ad*A(:,:, T).*(1-A(:,:,T)))+D(:,:,T);`	
		The time difference of D
33	`En=E(:,:,T)-N(:,:,T);En=En.*(En>0);`	Calculating En
34	`A(:,:,T+1)=dt*(ea*(1-A(:,:,T)).*En)+A (:,:,T);`	The time difference of A

(continued)

35	`end`	
36	`figure('Position',[0 100 1000 600]);`	
37	`for T=1:10:100`	Adjust the range of T for animation
38	`subplot(2,2,1);imagesc(E(:,:,T),` `[0 0.5]);axis xy;title('EGF');colorbar;`	
39	`subplot(2,2,2);imagesc(A(:,:,T),` `[0 1]);axis xy;title('AS-C');colorbar;`	

40	`subplot(2,2,3);imagesc(D(:,:,T),`
	`[0 0.5]);axis xy;title('Delta');`
	`colorbar;`	

41	subplot(2,2,4);imagesc(N(:,:,T),[0 0.2]);axis xy;title ('Notch');colorbar;	
42	pause(0.001);	
43	`end`	

In Fig. 4.48, the results for D and N are omitted, but you can see that they propagate with the proneural wave as in E. For more details of this mathematical model, please refer to other references [37–39].

4.9.4   Termination of the Wave in EGF Mutants

In the proneural wave experiments, it is important to perform **mosaic analysis** to study the effect of a cluster of cell mutants of a gene of interest. For example, EGF mutant cells do not produce EGF. But how can we simulate the effect of such mosaic regions? To do so, we only need to modify "script4_9B.m" slightly.

First, let's add "`Emutant=ones(Xmax,Xmax);Emutant(6:20,6:20)` `=0;`" in line 7 of "script4_9B.m" to give the distribution of EGF mutant cells as a matrix. In the region with a value of 1, EGF is produced, and in the region with a value of 0, it is not produced. Also, in Eq. (4.9.2), the only term that refers to the production of E is $a_e A(1 - A)$, and in "script4_9B.m," line 14, "`ae*A(:,:,T).*` `(1-A(:,:,T))`." We can multiply this term by the matrix "`Emutant`." Of course, this will be Hadamard product, so you can modify it as in "script4_9C.m" below:

script4_9C.m: modified from "script4_9B.m"

7	`Etemp=zeros(Xmax,Xmax);`
	`Eright=Etemp;Eleft=Etemp;`
	`Eup=Etemp;Edown=Etemp;`	

	`Emutant=ones(Xmax,Xmax);Emutant` `(6:20,6:20)=0;`	Definition of the EGF mutant region

(continued)

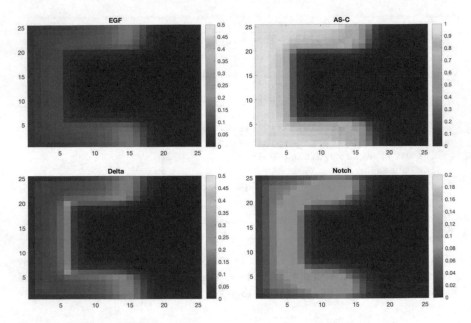

Fig. 4.49 Results of EGF mutant mosaic

| 14 | `E(:,:,T+1)=dt*(de*(Eright+Eleft +Eup+Edown-4*Etemp)/dx2` | |
| | `-ke*E(:,:,T)+ae*A(:,:,T).*(1-A (:,:,T)).*Emutant)+E(:,:,T);` | Reaction–diffusion of E in the presence of the EGF mutant region |

As shown in Fig. 4.49, not only E but also D, N, and A values do not increase in the EGF mutant region, and the wave of differentiation disappears.

4.9.5   Mechanism of Wave Acceleration in Notch Mutant

Now, we know experimentally that the proneural wave disappears in EGF mutant cells, which is the basic premise of this mathematical model. On the other hand, the question of why the differentiation wave is accelerated in Notch mutant cells has not been explained, because it is experimentally known that the wave progression is accelerated in Notch mutant regions, but the activity of EGF is lost at the same time [35]. EGF is an essential factor for wave progression, and if it is lost, the wave should also be lost. On the other hand, **why do the waves not disappear but rather accelerate in the Notch mutant region despite the loss of EGF?** This has been an unanswered question in the previous studies.

Now, let's modify "script4_9B.m" to simulate the Notch mutant region. Basically, it is the same as "script4_9C.m." First, let's add "Nmutant=ones(Xmax,

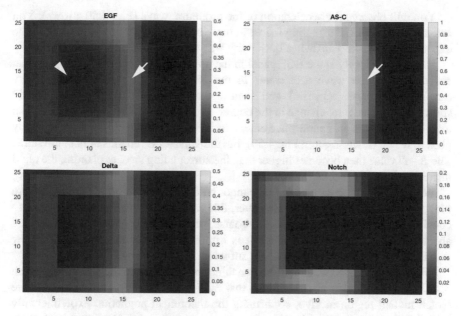

Fig. 4.50 Results of notch mutant mosaic

Xmax);Nmutant(6:20,6:20)=0;" in line 5 of "script4_9B.m" to set the
spatial pattern of the Notch mutation region. The term representing the production
of N in Eq. (4.9.4) is $d_n \sum_{l,m} D_{l,m} - d_c D_{i,j}$, which corresponds to "dn* (D (Y, X+1, T)
+D (Y, X-1, T) +D (Y+1, X, T) +D (Y-1, X, T)) -dc*D (Y, X, T)" in line 17 of
"script9B.m." All we have to do is to put all these terms related to the control of N in
parentheses and multiply by Nmutant. Note that the calculation here is not for the
whole matrix but for the elements one by one using the *for* statement. In other words,
it is the regular matrix product, "*Nmutant (Y, X)," not the Hadamard product.
The result will look like the following:

script4_9D.m: modified from "script4_9B.m"

5	N=zeros(Xmax,Xmax,Tmax);D=zeros (Xmax,Xmax,Tmax);	
	Nmutant=ones(Xmax,Xmax);Nmutant (6:20,6:20)=0;	Definition of the Notch mutant region, "Nmutant"
17	N(Y,X,T+1)=N(Y,X,T)+dt*((dn*(D (Y,X+1,T)+D(Y,X-1,T)+D(Y+1,X,T)	
	+D(Y-1,X,T))-dc*D(Y,X,T)) *Nmutant(Y,X)-kn*N(Y,X,T));	Calculating the lateral inhibition of N in the presence of Notch mutant region

Note that the Notch mutation region is assumed to exist only in the center of the
entire region, so there is no need to include "Nmutant" in the calculation for the
boundary regions such as X=1, 25 and Y=1, 25.

You will observe the acceleration of the proneural wave (Fig. 4.50 arrows). You can also see that EGF signaling is lost in the Notch mutant region, as shown in the experiment (Fig. 4.50 arrowhead). So why was the proneural wave accelerated?

If you look closely at the EGF panel in the upper left corner of Fig. 4.50, you can see that when the wave accelerates in the Notch mutant region, EGF activity transiently increases (arrows) and then decays (arrowhead). Equation (4.9.5) shows that the temporal change in the degree of cell differentiation A is proportional to $E - N$. When the proneural wave enters the Notch mutation region, it immediately goes to $N = 0$ (lower right in Fig. 4.50), but the value of E remains almost the same. The $E - N$ value then becomes higher than the surrounding area, explaining the rapid differentiation of the cells.

When there is a phenomenon that cannot be explained, we often think that there must be some unknown factor. However, this result means that we do not need to consider such unknown factors. Of course, this does not mean that there are absolutely no unknown factors, but it is important to note that the results show that EGF, Notch, Delta, and AS-C are sufficient.

When we examined whether the same thing that occurred in the above simulation results also occurred in vivo, we found that EGF activity transiently increased in the Notch mutant region in vivo, confirming the theoretical prediction experimentally [37]. In this way, problems in life science were actually solved by using mathematical models.

4.9.6   Why not a Salt-and-Pepper Pattern?

The above results indicate that Notch is indeed working and has the function of suppressing the progression of the proneural wave. But why does the lateral inhibition of Notch not produce the salt-and-pepper pattern shown in Fig. 4.41? This is the last question, so I will give you a hint and let you solve it by yourself.

As you can see in Fig. 4.41, if Delta were expressed at the same concentration in all cells, Notch signaling would be activated at the same magnitude in all cells, and lateral inhibition would be applied at the same magnitude. In this case, even if the lateral inhibition works, there will be no salt-and-pepper pattern. Even if the expression level of Delta appears to be uniform, subtle noise is added, and the noise fluctuation is amplified by lateral inhibition, resulting in a salt-and-pepper pattern. The expression level of Delta is not completely uniform in vivo because of the fluctuation of gene transcription, protein production, and signal transduction activity. It is not surprising that the salt-and-pepper pattern does not appear when no noise is added, so it is necessary to add noise in some way. Also, it is strange that noise is added only to Delta, so noise should be added to EGF, Notch, Delta, and AS-C.

We learned that we can get a random number using the *rand* command. For example, "rand(Xmax)" will give you an Xmax-by-Xmax matrix with uniform randomness in the range of 0 to 1. In general, we can add the random numbers as noise to the right-hand side of Eqs. (4.9.2), (4.9.3), (4.9.4), and (4.9.5), but actually, adding noise to the mathematical model of the proneural wave in this way causes a big problem. For example, if the value of E is zero, the right-hand side of Eq. (4.9.5) will always be zero, and A will never increase. However, if the value of E exceeds zero by even a small amount, such as 0.0001, then A will increase according to Eq. (4.9.5), and then E will increase according to Eq. (4.9.2). As a result, A will continue to increase until it reaches 1, and all cells will differentiate regardless of the progression of the proneural wave. This problem can be solved by considering another signaling pathway called Jak/Stat [39], but this would make the mathematical model more complicated, so let us consider a way to avoid the above problem without Jak/Stat.

Even if the value of E is originally zero, it will become a nonzero positive value when noise is added, but we can make it a multiplication instead of an addition. The right side of Eqs. (4.9.2), (4.9.3), (4.9.4), and (4.9.5) represents the change in EGF, Notch, Delta, and AS-C per unit time, and by multiplying all of them by a random number whose mean is 1, we can introduce noise so that the change in these values fluctuates within the range of plus or minus 10%. For example, in the matrix "0.9+0.2*rand(Xmax)," each matrix element will be a random value between 0.9 and 1.1 with 1 being the mean.

Exercise 4.9.6a Based on "script4_9B.m" introduce noise to the changes of EGF, Notch, Delta, and AS-C multiplying "0.9+0.2*rand(Xmax)."

Although *rand* produces uniformly distributed random numbers, it is unlikely that the noise added to gene expression or signaling activity would follow such a uniformly distributed random number. The simplest and easiest distribution of noise to consider is probably the **Gaussian normal distribution**. "randn(Xmax)" gives a Xmax-by-Xmax random matrix with mean 0 and standard deviation 1. If you want to set the standard deviation to 0.1, just set "1+0.1*randn(Xmax)" and multiply it by the amount of change in each variable. The 0.1 part can be replaced with a parameter for noise.

Exercise 4.9.6b Based on "script4_9B.m," make "script4_9E.m" by introducing noise to the changes of EGF, Notch, Delta, and AS-C multiplying "1+0.1*randn(Xmax)."

The result looks a little noisy, but I think the result is not so different (Fig. 4.51a). However, please note that this model contains a diffusible factor, EGF. Even if Notch is activated in a salt-and-pepper pattern as shown in Fig. 4.41, it is quite possible that this salt-and-pepper pattern is obscured by the diffusion effect of EGF.

If this is the case, why don't you reduce the amount of EGF produced in this noisy model? According to Eq. (4.9.2), a_e is the production rate of EGF, which was previously assumed to be $a_e = 1$.

Exercise 4.9.6c Set $a_e = 0.5$ in "script4_9E.m" to see what happens.

Now, you can see the salt-and-pepper pattern as shown in Fig. 4.51b. This result can be interpreted to mean that the lateral inhibition of Notch creates a salt-and-pepper pattern, but in the proneural wave model, its function is canceled by the

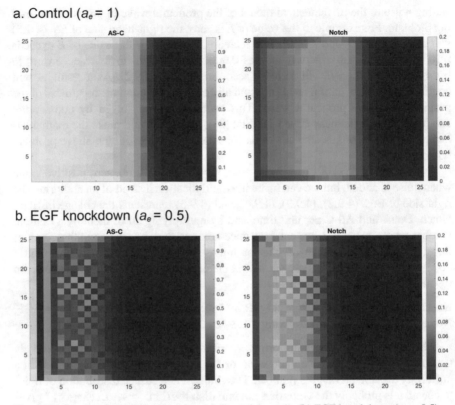

Fig. 4.51 Results of a model with noise. (**a**) Control ($a_e = 1$). (**b**) EGF knockdown ($a_e = 0.5$)

diffusible action of EGF. It is possible to experimentally reduce the production of EGF. However, if the production of EGF is completely eliminated, that is, $a_e = 0$, then nothing will happen because EGF is absolutely required for the proneural wave progression. The important thing here is to experimentally reproduce the operation corresponding to $a_e = 0.5$, not $a_e = 0$.

However, since each parameter of this mathematical model is an arbitrary value, we don't know what this value of 0.5 is exactly in vivo. Even if we knew, it would be technically difficult to accurately control the amount of EGF production in vivo.

In general, it is thought that eliminating a gene (**knockout**) results in complete loss of function of the gene product, but **knockdown** through RNAi or dominant-negative is thought to be less effective than knockout. In our paper [37], we found RNAi and dominant-negative fly strains that reduced EGF production to a good degree, and by using them, we found that Notch signaling and AS-C expression showed a salt-and-pepper pattern similar to that shown in the simulation.

Although it was known that the proneural wave disappears when EGF is completely knocked out [35], it was not known that a mild knockdown of EGF results in a salt-and-pepper pattern. Without the results of the computer simulation, we would not have bothered to conduct such an experiment. The results of this experiment, based on the predictions of the mathematical model, showed the correctness of the model. In addition, the interdisciplinary study revealed a new property of Notch signaling, in which **the function of Notch-mediated lateral inhibition is modulated to control the propagation speed of the proneural wave in the presence of the reaction–diffusion of EGF**. This is a good example of solving a biological problem using a mathematical model.

Note 10. Design of Equations

In this book, we have dealt with mathematical models of various life phenomena, but many of you may have wondered, "Why are such equations sufficient?" or "The actual phenomenon must be more complicated than that." For example, in Sect. 3.2 (Mathematical model of hematopoietic stem cells), Eqs. (3.2.1) and (3.2.2) do not include a logistic growth term. So, depending on the parameter settings, all cells will continue to grow indefinitely. However, this is not a problem if we focus on the initial period when the cells start to increase [5]. If we need to discuss the situation when the number of cells has stabilized over time, we can introduce the logistic term. In Sect. 4.7 (Turing model), we set upper limits on the concentration and production of diffusible substances, but mathematical studies on models with more complex reaction terms are known [17–20, 22]. Of course, it is quite possible that many unknown factors and unknown biochemical reactions will be included.

In Sect. 3.4 (Mathematical model of infectious disease), we discussed the most basic SIR model. However, a method based on the SIR model is also used in the simulation of COVID-19, which is currently a worldwide problem. For example, if you want to study the effect of the length of the incubation period, you can consider a four-variable model of $S, E, I,$ and R by adding the variable *Exposed* (E), which means that the patient is infected but not infectious (Exercise 3.4.1). Although the SIR model in this book does not consider spatial distribution, it can be extended to include human movement by combining it with the diffusion equation used in Chap. 4. If you want to study the effect of restricting the movement of people in some areas, you can use spatially inhomogeneous diffusion coefficients as in Sect. 4.8 (Simulation of heartbeats).

It is possible to develop a mathematical model that can represent all the processes rigorously, but biological phenomena are extremely complex, and it is a tremendous task to develop a mathematical model that can represent all the processes rigorously. For example, the diffusion equation describes the movement of molecules in a homogeneous medium by a random walk. However, in vivo, the diffusion of ligands is controlled by receptor binding and various glycoproteins, and the extracellular matrix may not be a homogeneous medium [4]. We do not yet have enough knowledge to accurately calculate the actual diffusion in vivo. Moreover, the more complicated the equation is,

(continued)

the more effort it takes to find the right parameters. Therefore, we will start with the simplest method of calculating the diffusion. In Sect. 4.6 (Pattern formation in fly leg), I explained that even a simple diffusion produces the concentric patterns due to the cooperative action of two morphogens.

Mathematical models are just that, models, and not reality. Therefore, it is important to consider how to "design" a mathematical model. For example, let's say you want to simulate a large number of people crossing a scramble intersection. Do we need to consider the thickness of each person's eyebrows at first? It is possible that the density of eyebrows is important, but it is not the first thing to consider. To simplify the model as much as possible, we can think of people as balls and replace the motion of people with that of many balls. Even this might yield some important insights. The next step is to replace the ball with a mannequin with arms and legs. This process of refining the model step by step is also important. If you can explain a phenomenon with the mannequin model that could not be explained with the ball model, you can understand the effect of having hands and feet. It is not always necessary to start with an elaborate model.

In Sect. 4.9 (Mathematical model of the wave of differentiation), Notch was activated only once in the proneural wave front (Fig. 4.43). However, Notch is actually activated again in neural stem cells, showing two peaks of activity [37]. Our recent work has shown that such two activity peaks can be reproduced with a slight modification of Eq. (4.9.4) [40]. By considering what kind of phenomenon a modification of the mathematical model corresponds to in vivo, we can expect to gain further insight into the function of Notch. In this way, we can modify the model little by little and compare the results.

Even though it is recommended to simplify the model, since it is a mathematical model of biological phenomena, it should be simplified so that it can be explained by biological knowledge as much as possible. This is because if each variable or parameter is linked to an actual gene or protein, the results obtained from the simulation can be verified experimentally. Various simplifications have been made in Sect. 4.9, but all equations can be explained biologically. In fact, by experimentally verifying the predictions of the mathematical model, the mechanism by which Notch-mediated lateral inhibition controls the wave progression was clarified. Thus, it will be important to design a mathematical model by fully utilizing biological knowledge.

Postscript

This book starts with the basics of programming using MATLAB and then explains how to solve ordinary and partial differential equations and perform various simulations based on mathematical models of life phenomena. While mathematical models using differential equations are common, our recent work has shown that geometrical models are effective in explaining tile patterns found in the compound eye of the fly. Further advances in the life sciences are expected by combining various mathematical methods with experimental biology [41].

There are situations where more advanced programming and mathematical knowledge are required. If this book is not enough for you, you can get more advanced knowledge from other books or the Internet. Since I myself am studying developmental biology using the fruit fly, many of the topics are related to flies, but I hope you have found that similar equations can be applied to a very wide range of life phenomena, and the methods described in this book can be applied to many different fields.

If you build a mathematical model of the phenomenon you are interested in by replacing the interaction of known factors with mathematical equations, and then run a computer simulation based on the model, you will hopefully be able to reproduce the actual phenomenon. And in that case, we can say that the known factors are sufficient to reproduce the phenomenon. However, I would like to see a further advancement. Wouldn't it be a surprise if some unexpected phenomenon is found from the simulations based on the mathematical model? Of course, we need to make sure that it is not a bug in the program, an inappropriate parameter, or a problem with the mathematical model itself, but if we can actually find the phenomenon in the simulation experimentally, it will not only confirm the correctness of the mathematical model, but also mean that **we have found a new mechanism of biological phenomenon. This will lead to the advancement of life science itself, beyond the simulation.** I hope that this book will help readers to make use of mathematical models and computer simulations in their studies and research in the life sciences.

M. Sato, *Getting Started in Mathematical Life Sciences*, Theoretical Biology,
https://doi.org/10.1007/978-981-19-8257-6

MATLAB Commands and Functions

The explanation here is quite simplified, so for details, please search the MATLAB documentation from the box with the magnifying glass in the upper right corner of the MATLAB window (Fig. 1.1).

sqrt: "sqrt(X)" calculates the square root of X (X can be a scalar, vector, or matrix).

abs: "abs(X)" calculates the absolute value of X (X is a scalar, vector, or matrix).

sin: "sin(X)" calculates the sine of X (X is a scalar, vector, or matrix).

cos: "cos(X)" calculates the cosine of X (X is a scalar, vector, or matrix).

tan: "tan(X)" calculates the tangent of X (X is a scalar, vector, or matrix).

asin: "asin(X)" calculates the inverse sine of X (X is a scalar, vector, or matrix).

acos: "acos(X)" calculates the inverse cosine of X (X is a scalar, vector, or matrix).

atan: "atan(X)" calculates the inverse tangent of X (X can be a scalar, vector, or matrix).

clear: Clear all variables in the workspace.

sum: "sum(X)" calculates the sum of all elements if X is a vector.
If X is a matrix, calculates the vector consisting of the sum of each column.
"sum(sum(X))" calculates the sum of all elements of a 2D matrix X.
"sum(X,'all')" calculates the sum of all elements of matrix X.

plot: Draw a 2D line graph.
For vectors X1, X2... and Y1, Y2....
"plot(X1, Y1)" plots a graph with X1 on the horizontal axis and Y1 on the vertical axis.
"plot(X1, Y1, X2, Y2)" plots graphs of X1 and Y1, and X2 and Y2 simultaneously.
You can have more than three pairs of vectors for the horizontal and vertical axes.
The number of elements in X1 and Y1, X2 and Y2... has to be the same.

Specify the style of each plot as "plot(X1,Y1,'style1',X2, Y2,'style2')".

The symbols used to specify the style are as follows:

Color: 'y' (yellow), 'm' (magenta), 'c' (cyan), 'r' (red), 'g' (green), 'b' (blue), 'w' (white), 'k' (black).

Line style: '-' (solid line), '--' (dashed line), ':' (dotted line), '-.' (chain line).

Point style: 'o' (circle), '+' (plus), '*' (asterisk), '.' (dot), 'x' (cross).

"plot(X1, Y1, 'r--o')" plots X1 and Y1 with red dashed lines and circles.

"plot(X,Y,'LineWidth',n)" sets the line thickness to n points.

title: Add a title to the graph (run immediately after *plot* or *imagesc*).

"title('arbitrary string')" adds a title to the last graph drawn.

xlabel: Add a label to the horizontal axis of the graph (immediately after *plot* or *imagesc*).

"xlabel('arbitrary string')" adds a label to the horizontal axis of the last graph drawn.

ylabel: Add a label to the vertical axis of the graph (run immediately after *plot* or *imagesc*).

"ylabel('arbitrary string')" adds a label to the vertical axis of the last graph drawn.

plot3: 3D version of *plot*.

X1, X2..., Y1, Y2..., and Z1, Z2.... are vectors.

The corresponding vectors have the same number of elements.

"plot3(X1, Y1, Z1)" plots a 3D graph using X1, Y1, and Z1.

"plot3(X1, Y1, Z1, X2, Y2, Z2)" plots two 3D graphs simultaneously.

zlabel: Add a label to the *z*-axis of the graph (run immediately after *plot3* or *surface*).

"zlabel('arbitrary string')" adds a label to the *z*-axis of the previously drawn graph.

zeros: Generate a matrix with all zero elements.

"zeros(n)" creates an n-by-n 2D matrix.

"zeros(n1, n2)" creates a 2D matrix with n1 rows and n2 columns.

"zeros(n1, n2, n3)" creates a n1-by-n2-by-n3 3D matrix.

ones: Generate a matrix with all elements 1.

"ones (n)" creates an n-by-n 2D matrix.

"ones (n1, n2)" creates a 2D matrix of n1 rows and n2 columns.

"ones (n1, n2, n3)" creates a n1-by-n2-by-n3 3D matrix.

rand: Generate a scalar, vector, or matrix whose elements are uniform random numbers of 0–1.

"rand" generates a random number (scalar) of 0–1.

"rand(n)" generates an n-by-n 2D matrix.

"rand(n1, n2)" generates an n1-by-n2 2D matrix.

"rand(n1, n2, n3)" generates a n1-by-n2-by-n3 3D matrix.

size: "size(X)" calculates the size of matrix X.

If X is an n-dimensional matrix, it returns the sizes of the all dimensions in the form of a vector.

"size(X, n)" returns the size of the n-th dimension of matrix X.

xlim: Set the range of the horizontal axis of the graph (run immediately after *plot* or *imagesc*).

"xlim([m n])" sets the range of the horizontal axis from m to n.

ylim: Set the range of the vertical axis of the graph (immediately after *plot* or *imagesc*).

"ylim([m n])" sets the range of the vertical axis from m to n.

randn: Generate a random number that follows a normal distribution (mean 0, standard deviation 1).

axis: Set the direction of the vertical axis and the ratio of the lengths of *x*, *y*, and *z* axes.

"axis xy" sets the direction of the vertical axis normal.

"axis equal" sets the unit length of *x*, *y*, and *z* axes to 1:1:1.

"axis square " sets the relative ratio of the total length along the *x*, *y*, and *z* axes to 1:1:1.

figure: Open a new window.

"figure('Position', [x y w h])" creates a new window with the leftmost coordinate

to x, the bottom coordinate to y, the width to w, and the height to h.

for/end/break: "for I=m:n" repeats the loop while changing I from m to n. Normally, I is incremented by 1, but in "for I=m:o:n" I is incremented by o. Multiple *for/end* statements can be combined to form multiple nested loops.

end returns to the most recently used *for* statement and iterates the loop.

break terminates the currently executing *for* statement loop.

fix: "fix(X)" truncate X to the decimal point (X can be a scalar, vector, or matrix).

ceil: "ceil(X)" round up X to the nearest whole number (X is a scalar, vector, or matrix).

round: "round(X)" rounds X to the nearest whole number (X is a scalar, vector, or matrix).

"round(X,n)" behaves same as "round(X)" if n=0.

If n>0, rounds X so that the right n digits of the decimal point remain.

If n<0, rounds X so that the left 1–n digits of the decimal point remain.

disp: "disp(X)" displays the variable X in the command window.

"disp('arbitrary string')" displays an arbitrary string enclosed in quotation marks (').

num2str: "num2str(X)" converts a numerical value X to a string.

strcat: "strcat(X, Y, Z)" generates a new string by concatenating strings X, Y, and Z.

randi: Generate scalars, vectors, and matrices consisting of uniformly random numbers of integers.

"randi(m)" generates 1integer random numbers between 1 and m (scalars).

"randi(m, n)" generates an n-by-n 2D matrix of integer random numbers between 1 and m.

"randi(m, n1, n2)" generates a n1-by-n2 2D matrix of random integers between 1 and m.

"randi (m, n1, n2, n3)" generates an n1-by-n2-by-n3 3D matrix of random integers between 1 and m.

while/end: Loop until *end* while the conditional expression after *while* holds.

"while I>m", if I is greater than m, repeat the loop.

"while I=<n", if I is less than or equal to n, repeat the loop.

"while (I<m) && (I>n)", if I is between n and m, repeat the loop.

"end" returns to the corresponding "while" statement and repeats the loop.

if/elseif/else/end: If the condition holds, execute the following instructions.

```
if condition 1          % if block
  · Instruction 1
elseif condition 2      % elseif block
     Instruction 2
else                    % else block
     Instruction 3
end
```

If condition 1 is satisfied, Instruction 1 is executed.

If not, and if condition 2 is satisfied, Instruction 2 is executed.

If not, Instruction 3 is executed.

"elseif" and "else" blocks can be omitted.

switch/case/end: Execute different instructions depending on the value following *switch*.

```
switch X
   case m            % case m block
      Instruction 1
   case n            % case n block
      Instruction 2
end
```

X can be a number or a string.

If X=m, Instruction 1 is executed.

If X=n, Instruction 2 is executed.

Number of "case" blocks can be changed.

function/end: Declare a new function

```
function Y=name (X1, X2)
   Instructions to calculate Y from X1 and X2
end
```

"name" is the name of the function, X1 and X2 are input variables, and Y is an output variable.

Number of input variables can be changed.

If there are multiple output variables, write them in the form of "[Y1, Y2, Y3...]".

imagesc: "imagesc (X)" displays the values of 2D matrix X.

"imagesc(X, [m n])" displays the 2D matrix X with the range of values from m to n.

Following *imagesc*, a color bar can be added by *colorbar*.

The display range of the horizontal and vertical axes can be set by *xlim* and *ylim*. The vertical axis direction can be set to normal by "axis xy".

colormap: "colormap(m)" changes the colormap to m.

The name of the colormap m can be selected from the following (you can also create your own):

parula: default colormap, blue for low values, yellow for maximum values.

jet: blue for low values, red for maximum values.

gray: black for low values, white for maximum values.

Various other colormaps are also available (see MATLAB documentation).

surface: "surface(X)" displays a 2D matrix X as a 3D image.

Draw a 3D image of the values of the elements of matrix X transformed to the Z-axis.

Click on the 3D rotation icon to rotate the image (Fig. 2.18).

repmat: Iterate over a matrix to generate a new matrix.

"repmat(X,m,n)" repeats the matrix X

m times in the row direction and n times in the column direction.

"repmat(X,m)" will repeat the matrix X m times in the row and column directions.

pause: "pause(m)" will stop the program execution for m seconds.

VideoWriter: Prepares to write a video file.

"m=VideoWriter(n)" creates an object m to save the file with file name n.

In "m=VideoWriter(n, p)", the video settings can be changed by p.

n and p are strings and need to be quoted.

If p is "MPEG-4", the video will be saved in MPEG-4 format.

See MATLAB documentation *VideoWriter* for details.

open: Open the file in the appropriate application.

Here, "open(m)" is used to open a video file specified by the object m generated by *VideoWriter*.

writeVideo: Write video data to a file.

"writeVideo(m,f)" writes video frame f to object m generated by *VideoWriter*.

f is retrieved by *getframe*.

getframe/gcf: Obtain an image as a video frame.

"getframe(f)" obtains the image of the graphics object f.

"getframe(gcf)" obtains the content of the last window drawn.

gcf: See *getframe/gcf*.

close: Close various files.

close(m) is used to close the object m created by *VideoWriter*.

legend: Add a legend to a graph (executed immediately after *plot or plot3*).

"legend('legend1', 'legend2',...)" adds a legend to the previous graph.

subplot: Create tiled coordinate axes.

"`subplot(m,n,p)`" divides the window into m rows and n columns and draws on the p-th tile.

Use it immediately before *plot* or *imagesc* to display images in a tiled format.

ode45: Compute numerical solutions of ordinary differential equations.

"`[T, Z]=ode45(f, [t1 t2], [x1; y1])`" computes the function f.

from time `t1` to `t2`, using initial values `x1` and `y1`.

The time is stored in vector form in T and the solution is stored in matrix form in Z.

Various instructions are available depending on the nature of the differential equation.

See the MATLAB documentation "Selecting an ODE Solver" for details.

global: Declare a global variable.

"`global m n X Y`" makes variables m, n, X, Y shared outside the function.

Comma (,) is not necessary.

Declare immediately after *function*.

Normally, each function is independent of the other functions inside the function set by *function*.

By making it a global variable, a variable with the same name can be shared by different functions.

It is necessary to declare a global variable in all target functions.

tic/toc: Measure the time taken between *tic* and *toc*.

Start the stopwatch at *tic*, stop at *toc*, and display the result.

toc: See *tic/toc*.

del2: Calculate the Laplacian.

"`del2(X, dx)`" computes 1/4 of the diffusion of a 2D matrix X.

dx is the spatial tick size of the derivative.

If dx=1, you can omit it and use "`del2(X)`".

References

1. Murray, J.D.: Mathematical Biology I. An Introduction, 3rd edn. Springer, Cham (2002)
2. Kreyszig, E.: Advanced Engineering Mathematics, 10th edn. Wiley, Hoboken (2011)
3. Smith, G.D.: Numerical Solution of Partial Differential Equations: Finite Difference Methods, 3rd edn. Oxford University Press, Oxford (1986)
4. Alberts, B.: Molecular Biology of the Cell, 6th edn. W.W. Norton, New York (2014)
5. Busch, K., Klapproth, K., Barile, M., Flossdorf, M., Holland-Letz, T., Schlenner, S.M., et al.: Fundamental properties of unperturbed haematopoiesis from stem cells in vivo. Nature. **518**(7540), 542–546 (2015)
6. Rodwell, V.W.: Harper's Illustrated Biochemistry, 30th edn. McGraw-Hill, New York (2015)
7. Kermack, W.O., McKendrick, A.G.: A contribution to the mathematical theory of epidemics. Proc. R. Soc. A. **115**(772), 700–721 (1927)
8. Lotka, A.J.: Elements of Physical Biology. Williams & Wilkins Co., Baltimore (1925)
9. Volterra, V.: Variazioni e fluttuazioni del numero d'individui in specie animali conviventi. Mem. Reale Accad. Naz. Lincei. **2**, 31–113 (1926)
10. Nellen, D., Burke, R., Struhl, G., Basler, K.: Direct and long-range action of a DPP morphogen gradient. Cell. **85**(3), 357–368 (1996)
11. Lecuit, T., Brook, W.J., Ng, M., Calleja, M., Sun, H., Cohen, S.M.: Two distinct mechanisms for long-range patterning by decapentaplegic in the drosophila wing. Nature. **381**(6581), 387–393 (1996)
12. Zecca, M., Basler, K., Struhl, G.: Direct and long-range action of a wingless morphogen gradient. Cell. **87**(5), 833–844 (1996)
13. Lecuit, T., Cohen, S.M.: Proximal-distal axis formation in the Drosophila leg. Nature. **388**(6638), 139–145 (1997)
14. Turing, A.M.: The chemical basis of morphogenesis. Philos. Trans. R. Soc. B. **237**(641), 37–72 (1952)
15. Hamada, H., Watanabe, M., Lau, H.E., Nishida, T., Hasegawa, T., Parichy, D.M., et al.: Involvement of Delta/Notch signaling in zebrafish adult pigment stripe patterning. Development. **141**(2), 318–324 (2014)
16. Inaba, M., Yamanaka, H., Kondo, S.: Pigment pattern formation by contact-dependent depolarization. Science. **335**(6069), 677 (2012)
17. Kondo, S., Asal, R.: A reaction-diffusion wave on the skin of the marine angelfish Pomacanthus. Nature. **376**(6543), 765–768 (1995)

18. Nakamasu, A., Takahashi, G., Kanbe, A., Kondo, S.: Interactions between zebrafish pigment cells responsible for the generation of Turing patterns. Proc. Natl. Acad. Sci. U. S. A. **106**(21), 8429–8434 (2009)
19. Shoji, H., Iwasa, Y., Kondo, S.: Stripes, spots, or reversed spots in two-dimensional Turing systems. J. Theor. Biol. **224**(3), 339–350 (2003)
20. Yamaguchi, M., Yoshimoto, E., Kondo, S.: Pattern regulation in the stripe of zebrafish suggests an underlying dynamic and autonomous mechanism. Proc. Natl. Acad. Sci. U. S. A. **104**(12), 4790–4793 (2007)
21. Yamanaka, H., Kondo, S.: In vitro analysis suggests that difference in cell movement during direct interaction can generate various pigment patterns in vivo. Proc. Natl. Acad. Sci. U. S. A. **111**(5), 1867–1872 (2014)
22. Meinhardt, H.: Models of Biological Pattern Formation. Academic Press, Cambridge (1982)
23. Watanabe, M., Kondo, S.: Is pigment patterning in fish skin determined by the Turing mechanism? Trends Genet. **31**(2), 88–96 (2015)
24. Kondo, S.: An updated kernel-based Turing model for studying the mechanisms of biological pattern formation. J. Theor. Biol. **414**, 120–127 (2017)
25. Gierer, A., Meinhardt, H.: A theory of biological pattern formation. Kybernetik. **12**(1), 30–39 (1972)
26. Alber, A.B., Paquet, E.R., Biserni, M., Naef, F., Suter, D.M.: Single live cell monitoring of protein turnover reveals intercellular variability and cell-cycle dependence of degradation rates. Mol. Cell. **71**(6), 1079–91.e9 (2018)
27. Hodgkin, A.L., Huxley, A.F.: A quantitative description of membrane current and its application to conduction and excitation in nerve. J. Physiol. **117**(4), 500–544 (1952)
28. FitzHugh, R.: Impulses and physiological states in theoretical models of nerve membrane. Biophys. J. **1**(6), 445–466 (1961)
29. Nagumo, J., Arimoto, S., Yoshizawa, S.: An active pulse transmission line simulating nerve axon. Proc. IRE. **50**, 2061–2070 (1962)
30. Courtemanche, M., Skaggs, W., Winfree, A.T.: Stable three-dimensional action potential circulation in the FitzHugh-Nagumo model. Physica D. **41**, 173–182 (1990)
31. Jongsma, H.J., Wilders, R.: Gap junctions in cardiovascular disease. Circ. Res. **86**(12), 1193–1197 (2000)
32. Danik, S.B., Liu, F., Zhang, J., Suk, H.J., Morley, G.E., Fishman, G.I., et al.: Modulation of cardiac gap junction expression and arrhythmic susceptibility. Circ. Res. **95**(10), 1035–1041 (2004)
33. Yu, S.R., Burkhardt, M., Nowak, M., Ries, J., Petrasek, Z., Scholpp, S., et al.: Fgf8 morphogen gradient forms by a source-sink mechanism with freely diffusing molecules. Nature. **461**(7263), 533–536 (2009)
34. Yasugi, T., Umetsu, D., Murakami, S., Sato, M., Tabata, T.: Drosophila optic lobe neuroblasts triggered by a wave of proneural gene expression that is negatively regulated by JAK/STAT. Development. **135**, 1471–1480 (2008)
35. Yasugi, T., Sugie, A., Umetsu, D., Tabata, T.: Coordinated sequential action of EGFR and Notch signaling pathways regulates proneural wave progression in the Drosophila optic lobe. Development. **137**(19), 3193–3203 (2010)
36. Sato, M., Suzuki, T., Nakai, Y.: Waves of differentiation in the fly visual system. Dev. Biol. **380**(1), 1–11 (2013)
37. Sato, M., Yasugi, T., Minami, Y., Miura, T., Nagayama, M.: Notch-mediated lateral inhibition regulates proneural wave propagation when combined with EGF-mediated reaction diffusion. Proc. Natl. Acad. Sci. U. S. A. **113**(35), E5153–E5162 (2016)

38. Sato, M., Yasugi, T.: Regulation of proneural wave propagation through a combination of notch-mediated lateral inhibition and EGF-mediated reaction diffusion. Adv. Exp. Med. Biol. **1218**, 77–91 (2020)
39. Tanaka, Y., Yasugi, T., Nagayama, M., Sato, M., Ei, S.I.: JAK/STAT guarantees robust neural stem cell differentiation by shutting off biological noise. Sci. Rep. **8**(1), 12484 (2018)
40. Wang, M., Han, X., Liu, C., Takayama, R., Yasugi, T., Ei, S.I., et al.: Intracellular trafficking of Notch orchestrates temporal dynamics of Notch activity in the fly brain. Nat. Commun. **12**(1), 2083 (2021)
41. Hayashi, T., Tomomizu, T., Sushida, T., Akiyama, M., Ei, S.I., Sato, M.: Tiling mechanisms of the *Drosophila* compound eye through geometrical tessellation. Curr. Biol. **32**(9), 2101–2109 (2022)

Index

© The Author(s), under exclusive license to Springer Nature Singapore Pte Ltd. 2022
M. Sato, *Getting Started in Mathematical Life Sciences*, Theoretical Biology,
https://doi.org/10.1007/978-981-19-8257-6